CONTENTS

BURMA IN THE AFTERMATH OF CYCLONE NARGIS: DEATH, DISPLACEMENT, AND HUMANITARIAN AID

HEARING

BEFORE THE

SUBCOMMITTEE ON ASIA, THE PACIFIC, AND THE GLOBAL ENVIRONMENT

OF THE

COMMITTEE ON FOREIGN AFFAIRS HOUSE OF REPRESENTATIVES

ONE HUNDRED TENTH CONGRESS

SECOND SESSION

MAY 20, 2008

Serial No. 110–229

Printed for the use of the Committee on Foreign Affairs

Available via the World Wide Web: http://www.foreignaffairs.house.gov/

U.S. GOVERNMENT PRINTING OFFICE

42–477PDF WASHINGTON : 2008

For sale by the Superintendent of Documents, U.S. Government Printing Office
Internet: bookstore.gpo.gov Phone: toll free (866) 512–1800; DC area (202) 512–1800
Fax: (202) 512–2104 Mail: Stop IDCC, Washington, DC 20402–0001

COMMITTEE ON FOREIGN AFFAIRS

HOWARD L. BERMAN, California, *Chairman*

GARY L. ACKERMAN, New York
ENI F.H. FALEOMAVAEGA, American Samoa
DONALD M. PAYNE, New Jersey
BRAD SHERMAN, California
ROBERT WEXLER, Florida
ELIOT L. ENGEL, New York
BILL DELAHUNT, Massachusetts
GREGORY W. MEEKS, New York
DIANE E. WATSON, California
ADAM SMITH, Washington
RUSS CARNAHAN, Missouri
JOHN S. TANNER, Tennessee
GENE GREEN, Texas
LYNN C. WOOLSEY, California
SHEILA JACKSON LEE, Texas
RUBÉN HINOJOSA, Texas
JOSEPH CROWLEY, New York
DAVID WU, Oregon
BRAD MILLER, North Carolina
LINDA T. SÁNCHEZ, California
DAVID SCOTT, Georgia
JIM COSTA, California
ALBIO SIRES, New Jersey
GABRIELLE GIFFORDS, Arizona
RON KLEIN, Florida
BARBARA LEE, California

ILEANA ROS-LEHTINEN, Florida
CHRISTOPHER H. SMITH, New Jersey
DAN BURTON, Indiana
ELTON GALLEGLY, California
DANA ROHRABACHER, California
DONALD A. MANZULLO, Illinois
EDWARD R. ROYCE, California
STEVE CHABOT, Ohio
THOMAS G. TANCREDO, Colorado
RON PAUL, Texas
JEFF FLAKE, Arizona
MIKE PENCE, Indiana
JOE WILSON, South Carolina
JOHN BOOZMAN, Arkansas
J. GRESHAM BARRETT, South Carolina
CONNIE MACK, Florida
JEFF FORTENBERRY, Nebraska
MICHAEL T. McCAUL, Texas
TED POE, Texas
BOB INGLIS, South Carolina
LUIS G. FORTUÑO, Puerto Rico
GUS BILIRAKIS, Florida
VACANT

ROBERT R. KING, *Staff Director*
YLEEM POBLETE, *Republican Staff Director*

SUBCOMMITTEE ON ASIA, THE PACIFIC, AND THE GLOBAL ENVIRONMENT

ENI F. H. FALEOMAVAEGA, American Samoa, *Chairman*

ADAM SMITH, Washington
GARY L. ACKERMAN, New York
GREGORY W. MEEKS, New York, *Vice Chair*
DIANE E. WATSON, California
RUBÉN HINOJOSA, Texas
ALBIO SIRES, New Jersey

DONALD A. MANZULLO, Illinois
DAN BURTON, Indiana
DANA ROHRABACHER, California
EDWARD R. ROYCE, California
STEVE CHABOT, Ohio
JEFF FLAKE, Arizona

LISA WILLIAMS, *Subcommittee Staff Director*
DAVID RICHMOND, *Subcommittee Professional Staff Member*
NIEN SU, *Republican Professional Staff Member*
VILI LEI, *Staff Associate*

BURMA IN THE AFTERMATH OF CYCLONE NARGIS: DEATH, DISPLACEMENT, AND HUMANITARIAN AID

TUESDAY, MAY 20, 2008

HOUSE OF REPRESENTATIVES,
SUBCOMMITTEE ON ASIA, THE PACIFIC,
AND THE GLOBAL ENVIRONMENT,
COMMITTEE ON FOREIGN AFFAIRS,
Washington, DC.

The subcommittee met, pursuant to notice, at 10:05 a.m., in room 2172, Rayburn House Office Building, Hon. Eni F.H. Faleomavaega (chairman of the subcommittee) presiding.

Mr. FALEOMAVAEGA. The subcommittee hearing will come to order.

This is the subcommittee hearing of the House Foreign Affairs Subcommittee on Asia, the Pacific, and the Global Environment. The topic for discussion of this oversight hearing is "Burma in the Aftermath of Cyclone Nargis: Death, Displacement and Humanitarian Aid" and the questions that are being raised by the members of this committee.

We have two distinguished members from the first panel—my good friend, the Deputy Assistant Secretary, Mr. Marciel, and also the Deputy Assistant Administrator, Mr. Gottlieb—that will be our first witnesses to testify this morning.

I will begin the hearing this morning with my opening statement. And I know my good friend, the ranking member of this subcommittee, the gentleman from Illinois, Mr. Manzullo, will be joining us later. I am happy to see that we have a distinguished member of our subcommittee here joining us, the Ambassador and distinguished colleague from California, Congresswoman Diane Watson, here with us.

On May 3rd of this year, Cyclone Nargis hit Burma, killing more than 100,000 people so far known. According to the United Nations, up to 2.5 million people in Burma may be affected due to the shortage of food, fuel, clean water and electricity.

While the United Nations' teams are on the ground, the military leaders are delaying aid and imposing travel restrictions on those providing assistance. Our own Congressional Research Service reports that, despite pledges of cash, supplies and assistance from around the world, most aid agencies have not been granted visas to enter Burma, or Myanmar, and there is still no word on when visas will be issued.

It is also reported that a relatively small number of foreign aid workers, mostly from Asian countries, will be allowed in. But it is not clear how far outside Rangoon they will be allowed to travel, because the military leaders insist that they can manage the relief effort and do not need experts.

As of May 15th, the international community had pledged more than $60 million in contributions and in-kind pledges. Japan and the United Kingdom have provided $10 million each, and the U.S. Agency for International Development, or USAID, reports that total humanitarian funding provided by the United States is $17.5 million to date.

United States airlifts of relief materials are being flown from Thailand to Rangoon. And I want to commend our Ambassador and my good friend, our Ambassador to Thailand, Ambassador Eric John, for the excellent work he is doing on behalf of the United States at this critical time.

I also want to express my disappointment, however, that the military leaders proceeded with a vote on this proposed Constitution at a time when the Burmese people, the people of Myanmar, are desperately trying to hold onto life and recover from the aftermath of Cyclone Nargis.

While voting was delayed for most of the townships around Rangoon and in several of the townships in Irrawaddy region, reports are swirling that the turnout for the rest of Burma was light and marred with voting irregularities, including premarked ballots, voter intimidation, et cetera. On the other hand, the State Peace and Development Council (SPDC) reports that more than 99 percent of eligible voters voted and that 92.4 percent voted in favor of the Constitution drafted by the SPDC.

On May 6th of this year, the U.S. House of Representatives passed House Concurrent Resolution 317 in hopes that the Burmese Government, or the Government of Myanmar, would call off the referendum and allow all resources to be focused on disaster relief to ease the pain and suffering of the Burmese people. I commend Congressman Joe Crowley from New York for offering this important resolution, of which I am an original co-sponsor.

While we remain on record in condemning the referendum going forward as scheduled, I am hopeful that the Burmese Government, or the Government of Myanmar, will at a minimum make disaster assistance a top priority.

I also hope that there might be more willingness on the part of the key nations, like China, India, Russia, and the ASEAN countries, to apply pressure on the SPDC or the military leaders that govern Myanmar to reduce human-rights abuses and begin a dialogue with Myanmar's opposition groups. To this day, Nobel Peace Prize recipient Aung San Suu Kyi remains under house arrest and Prime Minister Sein Win, who is with us today, remains in exile.

As an international community, we cannot turn a blind eye and pretend this is all right. To paraphrase Martin Luther King, Jr.'s statement, the people of Burma "will remember not the words of their enemies, but the silence of their friends."

So while I applaud the efforts of the United States and the international community in bringing aid to the people of Burma, I would hope you would also apply as much pressure on the military leaders of Myanmar to ensure a peaceful transition, hopefully, to democracy.

I also want to recognize the presence of another distinguished member of subcommittee, the gentleman also from California, Mr. Rohrabacher, if he has any opening statement.

[The prepared statement of Mr. Faleomavaega follows:]

PREPARED STATEMENT OF THE HONORABLE ENI F.H. FALEOMAVAEGA, A REPRESENTATIVE IN CONGRESS FROM AMERICAN SAMOA, AND CHAIRMAN, SUBCOMMITTEE ON ASIA, THE PACIFIC, AND THE GLOBAL ENVIRONMENT

On May 3, 2008, Cyclone Nargis hit Burma killing more than 100,000 people. According to the United Nations (UN), up to 2.5 million may be affected due to a shortage of food, fuel, clean water, and electricity.

While UN teams are on the ground, the military junta is delaying aid and imposing travel restrictions on those providing assistance. Our Congressional Research Service (CRS) reports that "despite pledges of cash, supplies, and assistance from around the world, most aid agencies have still not been granted visas to enter Burma and there is still no word on when visas might be issued."

It is also reported that a relatively small number of foreign aid workers, mostly from Asian countries, will be allowed in but it is not clear how far outside Rangoon they will be allowed to travel because the military junta insists that they can "manage the relief effort and do not need experts."

As of May 15, the international community had pledged more than $60 million in contributions and in-kind pledges. Japan and the United Kingdom have provided $10 million each, and the U.S. Agency for International Development (USAID) reports that total humanitarian funding provided by the U.S. is $17.5 million, to date.

U.S. airlifts of relief materials are being flown from Thailand to Rangoon, and I want to commend our U.S. Ambassador to Thailand, Eric John, for the excellent work he is doing on behalf of the U.S. at this critical time. I also want to express my disappointment that the military junta proceeded with a vote on its proposed constitution at a time when the Burmese people are desperately trying to hold on to life and recover from the aftermath of Cyclone Nargis.

While voting was delayed for most of the townships around Rangoon, and in seven of the townships in the Irrawaddy region, reports are swirling that the turnout for the rest of Burma was light and marred with voting irregularities including premarked ballots, voter intimidation, etc. On the other hand, the SPDC reports that more than 99 percent of eligible voters voted and that 92.4 percent voted in favor of the constitution drafted by the SPDC.

On May 6, 2008, the U.S. House of Representatives passed H. Con. Res. 317 in hopes that the Burmese government would call off the referendum and allow "all resources to be focused on disaster relief to ease the pain and suffering of the Burmese people." I commend Congressman Joe Crowley for offering this important Resolution, of which I am a co-sponsor. While we remain on record in condemning the referendum going forward as scheduled, I am hopeful that the Burmese government will, at a minimum, make disaster assistance a top priority.

I also hope there might be more willingness on the part of key nations such as China, India, Russia, and the ASEAN countries to apply pressure on the SPDC to reduce human rights abuses and begin a dialogue with Burma's opposition groups. To this day, Nobel Peace Prize recipient, Aung San Suu Kyi, remains under house arrest, and Prime Minister Sein Win, who is with us today, remains in exile. As an international community, we cannot turn a blind eye and pretend this is okay. To paraphrase Martin Luther King, the people of Burma "will remember not the words of [their] enemies but the silence of [their] friends."

So while I applaud the efforts of the U.S. and the international community in bringing aid to the people of Burma, I would hope we would also apply more pressure on the SPDC to ensure a peaceful transition to democracy.

I now recognize our Ranking Member for any opening statement he may have.

Mr. ROHRABACHER. Thank you very much, Mr. Chairman. And I see some old friends here. And just want to thank you very much for calling this hearing.

It is a timely hearing because the suffering of the people of Burma now is clearly evident. And I think we should all remember that this natural disaster that has struck Burma is second only to the disaster of their own repressive government, in terms of the quality of life and the horrible suffering that the people of Burma have had to endure.

Let us note that, throughout history, many of the dictators that held an iron grip on their people often were overthrown after a natural disaster, because it is usually in times of crisis, almost always in times of crisis, that people are able to see the fundamental nature of their own government.

Years ago, I remember people warning me that it was in times of crisis that the people of this country will determine who you are as a political leader, and it is not the everyday work that goes on which gives people the idea of just how we define ourselves through our own people and gives people an idea of what we stand for. It is in times of crisis when we define ourselves.

And in Burma, in this crisis, the military has defined itself as those of us who have opposed that dictatorship for so long expected. The dictatorship in Burma is a shameful, shameful institution that controls that country for the sake of controlling that country, for the sake of—the Government is operating for the sake of the people.

We should remember that Samosa, for example, was overthrown shortly after his disgraceful performance after a major earthquake in Nicaragua. Let us hope that this tragedy of the cyclone that has befallen the people of Burma can lead to something positive. Let us hope that the people of Burma, who, all along, have been trying to resist this dictatorship, that they will now succeed in creating a new Government.

Let us hope that the countries around Burma who have tolerated and been, perhaps, in coalition with those generals that have ruled Burma with an iron fist, let us hope that now that the Burmese dictatorship has performed in such an inhumane way, that these Governments will now back away at long last from any positive association with these gangsters that run that country.

It is said that there is "relief fatigue" settling into the people of United States, that we are becoming weary of helping people who are in crisis in times of natural disasters, like the cyclone that has hit Burma.

I would suggest that, while there is very reasonable question for the American people to say, "Why should we have the ongoing foreign aid programs that we do?" and to question at times, when some of our own people need help, why we have those ongoing relief efforts, well, that, I think, is reasonable. I think the American people do understand that emergency relief is different than long-term foreign aid; that, instead, emergency relief is what good people do to help others at a time when life-and-death issues are at stake because of, perhaps, natural disasters.

So I want to put myself on record as making sure that Republicans and Democrats understand that, when we help out people like the people of Burma who have suffered this cyclone, that it represents the best of the American character in times of emergency, even though we might debate over long-term foreign aid programs.

And I would hope that the people of Burma recognize that we, like we helped the people in Indonesia during that time of their tsunami, that this does reflect the heart and soul of America and that we are on their side.

And I would hope that the people of the military, at long last, understand that they should not be using their guns and their strength and their threatening power to keep in a position of authority those generals and gangsters who have controlled Burma for the last 40 years. This demonstration of the evil of the Burmese regime after this cyclone should convince the military that their guns are aimed in the wrong direction.

It is time for the Burmese military, from the junior officers on down, to say that the leadership of the Government of Burma no longer has any moral mandate from anyone to keep control of that country. And they should side with those who are calling for free elections and side with those democratic reformers, as represented by Aung San Suu Kyi.

And, finally, let us also note that China has played a very negative role in Burma. And China suffered—at the same time China has suffered, as now the people of Burma have suffered. And the Chinese stepped forward to help their people, unlike what the Government of Burma has done. So I would hope that the Chinese regime, itself, begins to distance itself from these gangsters who run that country of Burma.

And with that, Mr. Chairman, I thank you for holding this hearing. Those words and our commitment to providing emergency relief to people like the Burmese people who are suffering in such circumstances is something we need to restate, and this hearing will do so.

Mr. FALEOMAVAEGA. I thank the gentleman from California.

Now for the opening statement of Ms. Watson.

Ms. WATSON. Thank you, Mr. Chairman.

With the disaster in Myanmar, following on its heels the earthquake in China, I feel most of all that we may be witnessing disaster fatigue here in the United States and around the world before we have begun to respond in any meaningful way.

This is particularly true in Myanmar. Official figures from the May 3rd storm put the death toll at 78,000 and another 56,000 missing. Others estimate that the numbers are much larger. Two-point-four million people have been rendered homeless and in need of food. Whole areas in the Delta region are isolated, cut off and have not received assistance. It is difficult and mind-boggling to imagine so many people in need. I suppose if you imagine a large part of Metropolitan DC in such a state, we then may be able to put the disaster in perspective.

The Government of Myanmar has made an intolerable situation even more intolerable by refusing to open its doors to aid relief offered from nations around the world. Every day, lives are being lost due to the irresponsible and xenophobic reactions of a military dictatorship in Myanmar that maintains its presence through brutal suppression of the Burmese people.

What is particularly troubling is that the rice-planting season begins in June, which most likely has been interrupted by the cyclone and the influx of storm surge on arable land. I am told that, even before the storm, a significant number of children suffered from malnutrition.

I am also told that one-third of those killed by the cyclone are probably children. Tens of thousands of others who survived have lost their parents and now must suffer a child's worst nightmare. They live in a weakened state with vulnerable immune systems. It is they who will most likely die from water-borne diseases that many experts are expecting.

Mr. Chairman, as a former educator, I am particularly concerned about the plight of the children of Myanmar. When children lose all social structure, societies quickly fall apart. We saw it with the child soldiers in Sierra Leone and Liberia. It will take those societies generations of peace to recover. The situation in Myanmar is more ominous, given the fact that it has one of the world's largest rates of recruitment of child soldiers who are kidnapped and forced into joining the country's army.

The current disaster there has strained a bad situation for many of its own children. I would therefore be listening in this hearing to hear what our witnesses know about how international assistance efforts are being directed at this most vulnerable group and, more generally, what efforts are currently under way to break the logjam in assistance flowing into Myanmar.

Thank you so much for holding this very pertinent and timely hearing, Mr. Chairman.

Mr. FALEOMAVAEGA. I thank the gentlelady from California.

We are also joined this morning by another distinguished member of the House Foreign Affairs Committee, a colleague and dear friend who has been the chief sponsor of the recently passed resolution in addressing the problems and the concerns of the Congress on the issue of Myanmar, or, as some people call it, Burma, my good friend from New York, Mr. Crowley, if he has an opening statement.

Mr. CROWLEY. Mr. Chairman, firstly, let me thank you for once again showing your leadership on this issue, as you have been, I think, one of the foremost leaders on this issue from our side of the aisle. I want to thank you again for holding this timely hearing.

There is, I think, a tremendous humanitarian crisis that has been ongoing for a number of weeks now, as it pertains to the people of Burma. I am very, very interested in hearing from this distinguished panel before us today. We still don't know exactly how many people have died or been displaced, how many lives have been destroyed. However, it is being reported that tens of thousands that we know of have died. Hundreds of thousands have gone missing, and well over 1 million people have been displaced as a result of the cyclone.

It is not just a cyclone, though, that is causing the trauma to the people of Burma. I believe that the military junta has much to do with the anguish that many of the people of Burma are going through now.

Just yesterday, it was announced that Burma had agreed to let Southeast Asian neighbors help coordinate foreign relief effort. That is 2½ weeks after the cyclone first hit Burma. It has been suggested, and until I can be shown otherwise, that I believe possibly that crimes against humanity have occurred in Burma by the military junta in their neglect for their own people.

As was mentioned before, I introduced a resolution offering the condolences and sympathy of the people of the United States for the people of Burma and, once again, calling on the military junta to accept broad international assistance. In addition, I co-wrote a letter to the President to consult with the French, British and other supportive and regional Governments to coordinate a peaceful international humanitarian intervention and provide life-saving humanitarian aid to the victims of Cyclone Nargis.

Mr. Chairman, it is imperative that the people of Burma receive help immediately. I continue to hear reports that not only are the military junta refusing to grant visas for international aid for workers, including the USAID team of experts, it is also taking relief shipments and selling them back to the victims.

Burma has slowly permitted a small flow of aid from several nations, including the United States, but relief officials have said that this amounts to only 20 percent of the supplies that are needed. Without more aid, many, many more people will die of disease and starvation.

In reading The Washington Post yesterday, there was an op-ed talking about the need for food and water and hygiene, but that there is not the proper infrastructure for human waste disposal. That is just like a vicious cycle, that, unless we get additional assistance, practical assistance to the people of Burma, we are going to see more and more spread of deadly disease.

I believe we must do everything in our power to ensure that the 2.5-million-plus survivors in need of aid receive large-scale relief shipments, access to health care, et cetera, to meet the tremendous needs of the Burmese people.

I would also just make an observation, Mr. Chairman, before I turn it back to you and to our guest panelists, and that is the behavior of the junta, in my opinion, expresses a fear of international assistance because they believe that the assistance will turn into a revolution against their Government.

That is, I think, right now what is least on the minds of people who want to help in this situation. This is a humanitarian crisis. We are trying to get food and aid to people who are suffering, the children who are suffering and have lost their parents, psychological help to the victims in Burma. But it also says to me that they know, they know deep down inside, that what they are doing is wrong, that they are morally corrupt, in that they put the interest of the junta and their so-called Government above the people themselves.

And so, with that, Mr. Chairman, I would yield back and look forward to hearing the testimony.

Mr. FALEOMAVAEGA. I thank the gentleman from New York for his comment and statement on this hearing.

And we are going to do all we can to see what we need to do. Hopefully, after the hearing, there may be some offered suggestions or recommendations from our good friends downtown that we might be able to get a better picture of what we need to do on the part of Congress.

As I said earlier, we are joined here this morning by the members of our first panel.

My good friend, Deputy Assistant Secretary of the Bureau of East Asian Affairs, Pacific Affairs, Mr. Scot Marciel. Mr. Marciel is a career Foreign Service Officer; has served posts in Vietnam, the Philippines, Hong Kong, Brazil and Turkey, and recently also was appointed to the Economic Bureau's Office of Monetary Affairs with the State Department. A graduate of the University of California at Davis and also from the Fletcher School of Law and Diplomacy.

Also with us is the Deputy Assistant Administrator of USAID'S Bureau for Democracy, Conflict and Humanitarian Assistance, Mr. Gregory Gottlieb. He has also served at posts in Kenya and Sudan and Ethiopia and Malawi. Mr. Gottleib has had more than 25 years of experience in humanitarian assistance with the United States Government and as well as with the NGOs and the United Nations. A law graduate from the Loyola Law School and also received his master's degree with the Harvard University Kennedy School of Government.

Gentlemen, I welcome you both to testify this morning. By unanimous consent, your statements will be made part of the record, and any other additional materials that you want to add on to it, it will be made part of the record.

Secretary Marciel, your statement.

STATEMENT OF THE HONORABLE SCOT MARCIEL, DEPUTY ASSISTANT SECRETARY, BUREAU OF EAST ASIAN AND PACIFIC AFFAIRS, U.S. DEPARTMENT OF STATE

Mr. MARCIEL. Mr. Chairman, Ranking Member Manzullo and members of the subcommittee, thank you for inviting me here today to testify about the ongoing humanitarian crisis in Burma.

On May 2nd and 3rd, Tropical Cyclone Nargis struck Burma with 130-mile-per-hour winds and torrential rain. The U.N. estimates cyclone-related deaths at between 78,000 and 125,000, with 2.4 million people left in urgent need of assistance. More than 95 percent of structures in some communities in the Irrawaddy Delta, the region hit hardest by the cyclone, were demolished.

After the cyclone struck Burma, the United States responded rapidly to offer its assistance to the people affected by the storm. My colleagues from USAID will speak in detail to this issue. But I will note that, to date, we have made available more than $17.5 million in humanitarian assistance.

On May 12th, Director for Foreign Assistance Henrietta Fore, Commander of U.S. Pacific Command Admiral Timothy Keating and I accompanied the first C–130 carrying United States relief supplies into Burma. Meeting with the senior Burmese military delegation, we explained our offer of assistance and conveyed our desire to help avoid needless loss of life in the wake of this natural disaster.

We have not been alone in offering help. The international community has responded by offering over $100 million in assistance to Burma. Many nations have offered to send doctors and disaster relief teams.

Sadly, the Burmese regime's response to this disaster has fallen far short of what was required. Frankly, it has been appalling. Rather than dedicating its full attention to the humanitarian disaster inflicted by the cyclone, the Burmese regime gave priority to conducting a deeply flawed referendum on its draft Constitution on May 10th in all states and divisions outside of those declared disaster areas. The fact that the regime proceeded with the referendum in the midst of this humanitarian disaster speaks volumes to its indifference to the welfare of the Burmese people.

Even more disturbing has been the regime's refusal, to date, to accept offers of outside expertise, its failure to allow international relief workers access to the affected areas to assess the situation and carry out relief work, and its insistence that it is capable of managing the logistics of the aid-distribution operation. It clearly is not. Critical shortages abound: Helicopters and pilots that carry supplies to inaccessible areas; doctors to treat the sick and prevent infection; and public health experts to provide sanitation facilities.

We have seen some positive news in recent days, such as an increase in the flow of relief goods into the Delta region. The regime apparently also has agreed to grant visas to health workers from neighboring countries, and some additional U.N. and European aid personnel have gotten in. International NGOs and U.N. relief agencies seem to be expanding their footprint in the affected areas.

Still, the situation is increasingly desperate. More than 2 weeks after the cyclone hit, the U.N. estimates that only one in four people have been reached. The door must be opened far wider, and rapidly, to prevent a second catastrophe.

If assistance and expertise is not allowed in and thousands of Burmese perish, the responsibility for this catastrophe will fall squarely on the shoulders of Senior General Than Shwe and the other Burmese leaders. We call on those leaders to work with their neighbors and the broader international community to help save lives by accepting the offers of logistical support and technical expertise and by allowing full, unfettered access to the affected areas.

The United States Government is engaged in intensive diplomatic efforts to convince the regime to accept the outside assistance that the people of Burma so desperately need. President Bush spoke with Chinese President Hu on May 13th, and Secretary Rice has reached out to her counterparts, including in China and India, as have many other senior officials.

Others have pressed a similar message, including European nations, China, India, and the ASEAN countries, as well, of course, as the United Nations. We welcome ASEAN's convening of an emergency foreign ministers' meeting yesterday on this subject and hope that the proposed ASEAN assistance mechanism announced at that meeting will succeed in rapidly accelerating the flow of assistance.

We, likewise, hope that the pledging conference that U.N. Secretary-General Ban and ASEAN will co-host in Rangoon on May 25 will lead to greater access for international relief workers and supplies.

We will continue to exhaust all diplomatic channels and opportunities to persuade the regime to grant access to the experts and assets that can expedite the flow of humanitarian assistance to those in need.

Given the regime's resistance to grant access for international relief teams to the affected areas of the Delta, our initial contributions of relief supplies were handed over to Burmese authorities for transportation to the disaster areas. We have made every possible effort to monitor the disposition of those supplies. And since May 16th, we have been able to consign the contents of two relief flights daily directly to international NGOs.

While the United States is focused at the moment on preventing further needless loss of life, our fundamental policy toward the Burmese regime has not changed. We continue to maintain that the regime's referendum was not free, fair, nor credible. The regime's claims that 99 percent of eligible voters turned out on May 10th and that the 92 percent of voters supported the draft Constitution lack any credibility.

The United States, along with many in the international community, remains convinced that the only hope for a peaceful transition to a genuinely democratic government in Burma is through a substantive, time-bound dialogue between the regime and Burma's democratic and ethnic minority opposition.

An important first step toward a transition to democracy would be the release of all political prisoners. The most recent extension of Aung San Suu Kyi's house arrest is set to expire this weekend. We renew our call for the release of Aung San Suu Kyi and all other political prisoners.

We remain mindful of the need to address the crisis of governance in Burma while we continue to do all we can to convince the regime to permit the international community to provide the help that is so desperately needed in the wake of this humanitarian crisis.

Thank you for extending this opportunity for me to testify. I welcome your questions.

[The prepared statement of Mr. Marciel follows:]

PREPARED STATEMENT OF THE HONORABLE SCOT MARCIEL, DEPUTY ASSISTANT SECRETARY, BUREAU OF EAST ASIAN AND PACIFIC AFFAIRS, U.S. DEPARTMENT OF STATE

Mr. Chairman, Ranking Member Manzullo, and Members of the Subcommittee, thank you for inviting me here today to testify about the ongoing humanitarian crisis in Burma and our efforts to assist the victims of Tropical Cyclone Nargis.

On May 2–3, Tropical Cyclone Nargis struck Burma with 130 mile per hour winds and torrential rain. More than 95 percent of structures in some communities in the Irrawaddy Delta, the region hit hardest by the cyclone, were demolished in the storm. Twelve foot storm surges inundated countless villages. Downed trees and power lines and washed-out roads and bridges complicated transportation and communication in Rangoon and beyond. The storm left 2.4 million in urgent need of assistance. The UN estimates cyclone-related deaths at between 78,000 and 125,000. The suffering of the victims of this natural disaster has been compounded by continuing bad weather and heavy rains.

After Cyclone Nargis struck Burma, the United States responded rapidly to offer its assistance to the people affected by the storm. My colleague from USAID will speak in detail to this issue, but I will note that to date USAID has made available more than $17.5 million in humanitarian assistance, over $16 million of which has gone directly to United Nations programs and trusted non-governmental organizations. We have also offered a Disaster Assistance Response Team and military assets to augment the Burmese regime's limited capacity to provide disaster relief. Defense has provided additional commodities. On May 12, Director for Foreign Assistance Henrietta Fore, the Commander of U.S. Pacific Command, Admiral Timothy Keating, and I accompanied the first C–130 carrying U.S. relief supplies into Burma. Meeting with a senior Burmese military delegation, we explained our offer of technical and logistical assistance and conveyed our desire to help avoid needless loss of life in the wake of this natural disaster.

The United States is not alone in offering assistance. The international community has responded by offering over $100 million in assistance to Burma. Many nations have offered to send doctors and disaster relief teams. Non-governmental organizations and the private sector also have responded generously with cash and in-kind assistance.

Sadly, the Burmese regime's response to this disaster has fallen far short of what was required. Frankly, it has been appalling. Rather than dedicating its full attention to the humanitarian disaster inflicted by Cyclone Nargis, the Burmese regime gave priority to conducting a deeply flawed referendum on its draft constitution on May 10 in all states and divisions outside of those declared disaster areas in the wake of the storm. The fact that the regime proceeded with the referendum in the midst of a humanitarian disaster speaks volumes to its indifference to the welfare of the Burmese people.

Even more disturbing has been the regime's refusal to date to accept offers of outside expertise and its insistence that it is capable of managing the logistics of the aid distribution operation. It clearly is not. Critical shortages abound—helicopters and helicopter pilots to ferry supplies to inaccessible areas; doctors to treat the sick and prevent infection; and public health experts to provide sanitation facilities. We and the international community remain ready to provide this kind of help, as we did after the 2004 tsunami and as we are doing now for China in the aftermath of its devastating earthquake last week.

We have seen some positive news in recent days, such as an increase in the flow of relief goods into the Delta region. The regime has also apparently agreed to grant visas to health workers from neighboring countries and some additional UN and European aid personnel have gotten in. International NGOs and UN relief agencies seem to be expanding their footprint in the affected areas. The regime said it would grant blanket flight clearances for relief flights, and transport networks are improving slowly, with the bulk of supplies being transported by truck and boat to the affected area.

Still, the situation is increasingly desperate, and the regime's failure to provide greater access for the international community to the affected area is putting hundreds of thousands of lives at risk. More than two weeks after the cyclone hit, the UN estimates that, at best, only one in four people have been reached. The door must be opened far wider—and rapidly—to prevent a second catastrophe. Let me be clear: if assistance is not allowed in, and thousands of Burmese perish, the responsibility for this catastrophe will fall squarely on the shoulders of Senior General Than Shwe and other Burmese leaders. We call on those leaders to work with their neighbors and the broader international community to help save lives by accepting the offers of logistical support and technical expertise, and by allowing full, unfettered access to the affected areas.

The United States Government has engaged in intensive diplomatic efforts over the past two weeks to convince the regime to accept the outside assistance that the people of Burma desperately need. President Bush spoke with Chinese President Hu on May 13, and Secretary Rice has reached out to her counterparts, including in China and India. Deputy Secretary Negroponte raised this issue during his recent travel to South Korea, Japan, and China. Secretary of Health and Human Services Leavitt discussed Burma with his counterpart in China and other senior government officials during his trip to Beijing last week, and has held two conversations on the matter in recent days with the Director-General of the World Health Organization. Our Chargé in Rangoon, Shari Villarosa, has engaged directly with senior representatives of the regime as well. I was in the region when the storm struck and took advantage of my meetings there, including at the ASEAN Regional Forum Senior Officials Meeting and the U.S.-ASEAN Dialogue to reiterate our call for urgent access for humanitarian aid experts. Our ambassadors throughout the region and beyond echoed this message in conversations with senior government officials.

Others have pressed a similar message. China, India and the ASEAN countries have responded, some helpfully sharing their own experiences with humanitarian disaster relief. ASEAN Member States, including Burma, convened in Singapore on Monday to discuss disaster relief efforts and the establishment of a possible UN–ASEAN coordination mechanism.

We are also actively engaging in the UN. Secretary-General Ban and Under Secretary General for Humanitarian Affairs John Holmes have spoken out strongly on this issue. Burmese authorities over the weekend invited UN Secretary General Ban to Burma and he has accepted. We hope this is the beginning of a new approach by the regime. Separately, France has suggested that the UN Security Council invoke the "responsibility to protect" principle, to authorize international humanitarian assistance efforts if the Burmese regime continues its refusal to grant access. We will continue to exhaust all diplomatic channels and opportunities to persuade the regime to grant access to the experts and assets that can expedite the flow of humanitarian assistance to those in need.

The UN and ASEAN have announced plans to host a pledging conference in Rangoon on May 25, which UN Secretary-General Ban will attend. We are now reviewing possible U.S. participation in this conference. However, we still believe that the key to saving more lives is to increase access urgently to the disaster areas for international relief teams who can provide the expertise and logistical resources that the Burmese regime lacks. Without an adequate and independent assessment of the situation and current needs, as well as a commitment by the regime to provide the necessary access, a pledging conference is unlikely to produce the results we seek.

Given the regime's resistance to grant access for international relief teams to the affected areas of the Delta, our initial contributions of relief supplies were handed over to Burmese authorities for transportation to the disaster areas. Recognizing that this is a calculated risk, and given the regime's track record, we have made every possible effort to monitor the disposition of those supplies. U.S. Embassy staff are stationed at the airport in Rangoon to monitor the offloading of relief supplies. The Embassy is in frequent contact with UN and non-governmental partners operating in the disaster areas and also monitors markets in Rangoon and outside the city for indications that relief supplies are being diverted and sold. So far, we have been unable to confirm any such reports. Since May 16, we have been able to consign the contents of two relief flights daily directly to international NGOs, and hope we will be able to increasingly operate that way in the future. Our UN and NGO partners indicate that relief supplies are increasingly getting through to the disaster areas, although much more is still needed.

While the United States is focused at the moment on preventing further needless loss of life, our fundamental policy toward the Burmese regime and its self-described "roadmap to democracy" has not changed. We continue to maintain that the regime's referendum has not been free, fair, or credible. The regime's claims that 99 percent of eligible voters turned out on May 10, and that 92 percent of voters supported the draft constitution, lack any credibility. These results obviously cannot be considered representative of the will of the Burmese people.

The United States, along with many in the international community, remains convinced that the only hope for a peaceful transition to a genuinely democratic government in Burma is through a substantive, time-bound dialogue between the Burmese regime and Aung San Suu Kyi and other democratic and ethnic minority leaders. Promoting such a dialogue and transition to democratic rule remains our primary, over-arching foreign policy objective in Burma.

At this moment, however, we are focused on the humanitarian disaster and doing whatever we can to minimize the needless loss of life in the wake of this terrible tragedy, while recognizing that the best hope for a successful intervention lies in a decision by the Burmese regime to permit the international community to provide the help that is so desperately needed.

Thank you for extending this opportunity to me to testify today on these most serious issues in Burma. I welcome your questions.

Mr. FALEOMAVAEGA. Thank you, Mr. Secretary.

Mr. Gottlieb?

STATEMENT OF MR. GREG GOTTLIEB, DEPUTY ASSISTANT ADMINISTRATOR, BUREAU FOR DEMOCRACY, CONFLICT, AND HUMANITARIAN ASSISTANCE, U.S. AGENCY FOR INTERNATIONAL DEVELOPMENT

Mr. GOTTLIEB. Thank you, Mr. Chairman and members of the subcommittee, for the opportunity to appear before you today to discuss the humanitarian emergency in Burma.

Tropical Storm Nargis made landfall at about 1600 hours local time on Friday, May 2nd, in the Irrawaddy River Delta. At landfall, Nargis had winds of approximately 132-miles-per-hour and produced a storm surge of 12 feet. To help illustrate how strong that is, I note that Hurricane Katrina had sustained winds of 125-miles-per-hour when it made landfall. The subsequent storm surge, rain and wind devastated the low-lying Delta region and the farmers and fishermen who live there.

The magnitude of the death and suffering is overwhelming. The U.N. reports that cyclone-related deaths could range from 63,000 to more than 101,000. An additional 220,000 people are still believed to be missing. However, the Burmese Government estimates are lower. On state-run TV, the Government of Burma has confirmed 77,000 deaths but only 56,000 missing persons as of May 19.

Aid agencies working in the region report that hundreds of thousands of people are now homeless. Some are reportedly moving from place to place in search of the very basics: Food, water, medical care and shelter. At least 150,000 people are displaced in a mixture of 120 official and unofficial temporary settlements with water and sanitation conditions reportedly highly inadequate.

The Government of Burma has been slow to grant visas for international aid workers, including for USAID's team of experts. Those aid workers who receive visas are mostly confined to Rangoon City, leaving Burmese staff to carry on the enormous task of getting aid to all those in need. However, the poor and damaged infrastructure and the limited humanitarian capacity on the ground are preventing the aid community from fully understanding who needs what and where.

We have repeatedly emphasized to the Burmese Government the urgent need to allow humanitarian aid workers access to undertake thorough assessments.

Our reports indicate that supplies are slowly getting out to affected areas. At best, 20 percent of the 2.4 million people who are affected had received any aid as of May 19. Only .25 million of those 2.4 million people have received food through the World Food Programme. Our own 20 relief flights of USAID commodities completed between May 12th and 19th will serve more than 106,000 people with merely a fraction of the supplies they need.

I would like to explain in further detail the extent of the aid that is required to help those affected by the cyclone. The U.N. Office for the Coordination of Humanitarian Affairs reports that critical needs include water, sanitation, food, shelter and medical care. Let me highlight two concerns.

No disease outbreaks have been reported to date, but the U.N. World Health Organization has reported that an increase in the area in which mosquitoes and other insects can breed is expected to result in higher-than-usual numbers of people becoming sick with malaria and dengue fever within weeks. Even during an average rainy season in the Delta, incidence of these illnesses is high. If medical staff cannot get to those in need to provide preventive and curative care, the outcome may be dire.

Food security is also a major concern for relief organizations. Even in good times, the nutritional status of those living in the Delta region is poor. The U.N. notes that in cyclone-affected areas more than 30 percent of children under 5 years old are chronically malnourished.

As of May 18th, the U.N. World Food Program (WFP) had dispatched more than 1,200 metric tons of rice, high-energy biscuits and beans to Burma's cyclone-affected areas. However, WFP will need to move 390 metric tons of food every day if it is to reach the .75 million people it is targeting over the next 30 days.

The USAID response to the crisis in Burma has three components: Technical assistance, relief supplies, and funding through the United Nations and nongovernmental organizations.

Our DART team is our first component. The team remains in Bangkok, Thailand, where they are coordinating with DoD, U.N. and NGO partners, but still are without access to affected populations in Burma.

When a Disaster Assistance Response Team is not present during a large-scale emergency we are not able to adequately assess and coordinate efforts with our international counterparts. We cannot talk to those who are affected to see damage firsthand. It takes us longer to determine the best course of action and use of our resources.

USAID Administrator Henrietta Fore, Admiral Tim Keating, commander of PACOM, Scot Marciel, and our DART team leader, Bill Berger, went in on the first airlift of U.S. supplies on May 12th. Those supplies were handed over to the Burmese authorities, who had helicopters waiting to move them.

The U.S. Embassy staff members have been monitoring the off-loading of U.S. relief supplies on subsequent flights and, as of May 15th, had not found any verification of the rumors that USAID supplies had been diverted. Others on the ground also have reported that they have not witnessed our supplies being sold in local markets or diverted.

On May 16th, USAID began handing over supplies directly to nongovernmental organizations which had the ability to transport supplies to the affected areas and distribute them to beneficiaries. This shift will provide more accountability to ensure that United States supplies are getting to the Burmese people.

The third component is the funding we provide to NGOs and U.N. organizations that are working in the affected areas. As of May 19th, the U.S. Agency for International Development (USAID) has provided $17.5 million in aid to Burma. Since May 5th, we have provided $5.5 million in commodities and funding to U.N. agencies, the Red Cross and NGO partners for health, logistics, shelter, water sanitation and hygiene. And I know from Ms. Watson that there are also programs through Save the Children and UNICEF targeted at children and mothers.

In addition, USAID's Office of Food for Peace is providing $12 million of P.L. 480 Title II Assistance through the World Food Program. When combined with the cost of relief supplies and funds from the Department of Defense, this brings total U.S. Government humanitarian assistance to more than $19 million.

Historical experience shows that even the best-prepared country facing a disaster of this magnitude requires international assistance. USAID is working with its partners to ensure that relief gets to those who need it. We are closely monitoring reports that would indicate deterioration in the health situation, and we are working to ensure relief supplies and services reach those who need them.

Above all, I want to emphasize that USAID remains ready to help. We continue to stand by the people of Burma during this difficult time.

Thank you, Mr. Chairman, for taking time to discuss this important issue. I would be happy to answer any questions you may have.

[The prepared statement of Mr. Gottlieb follows:]

PREPARED STATEMENT OF MR. GREG GOTTLIEB, DEPUTY ASSISTANT ADMINISTRATOR, BUREAU FOR DEMOCRACY, CONFLICT, AND HUMANITARIAN ASSISTANCE, U.S. AGENCY FOR INTERNATIONAL DEVELOPMENT

BURMA IN THE AFTERMATH OF CYCLONE NARGIS

Thank you Mr. Chairman and Members of the subcommittee for the opportunity to appear before you today to discuss the humanitarian emergency in Burma.

I want to begin by extending the deepest condolences to the people of Burma. The outpouring of concern for Burma has been overwhelming here in the United States. I want to assure them that the American people are doing everything they can to help the Burmese people in their hour of need. And the need is immense and immediate.

CURRENT SITUATION

Tropical Storm Nargis made landfall at about 1600 hours local time on Friday, May 2, to the west of Rangoon in the Ayeyarwady River delta. At landfall, Nargis had winds of approximately 132 mph and produced a storm surge of 12 feet—roughly equivalent to a Category Four hurricane. To help illustrate how strong that is, I note that Hurricane Katrina had sustained winds of 125 mph when it made landfall in Louisiana. Although Nargis was downgraded at landfall, the subsequent storm surge, rain, and wind devastated the low-lying delta region and the farmers and fishermen who live there. The cyclone then continued east-northeast and caused extensive damage in the city of Rangoon.

The magnitude of the death and suffering is overwhelming. As of May 19, the U.N. estimated that 2.4 million people were affected. Cyclone-related deaths could range from 63,000 to more than 101,000, the U.N. reports. An additional 220,000 people were still believed to be missing. However, the Burmese government estimates are lower. On state-run TV, the GOB has confirmed 77,000 deaths and 56,000 missing persons, as of May 19.

Those who survived are suffering immensely. Aid agencies working in the region report that hundreds of thousands of people are now homeless. Some are reportedly moving from place to place in search of the very basics—food, water, medical care for injuries sustained during the cyclone, and plastic sheeting to protect them from continuing seasonal rains. At least 150,000 people are displaced in a mixture of 120 official and unofficial temporary settlements in the Ayeyarwady Delta, as of May 18. What we understand about these gatherings is that many are overcrowded, and that sanitation conditions are appalling.

Unfortunately, despite the tireless efforts of the aid personnel on the ground, the international humanitarian community still does not have a complete and accurate picture of the situation. The Government of Burma has been slow to grant visas for international aid workers, including for USAID's team of experts. Those aid workers who receive visas are mostly confined to Rangoon city. This leaves Burmese staff—many of whom themselves lost their homes and loved ones—to carry on the enormous task of getting aid to all those in need. Their efforts are extraordinary. However, the poor and damaged infrastructure and the limited humanitarian capacity on the ground are preventing the aid community from fully understanding who needs what, and where. This is why at all levels of the U.S. Government we have repeatedly emphasized to the Burmese government the urgent need to allow humanitarian aid workers access to do what they are trained to do in these situations: quickly assess the needs and recommend the best course of action to save lives and alleviate suffering.

On May 18, U.N. Emergency Relief Coordinator John Homes arrived in Burma to meet with Burmese government officials and visit the severely affected Ayeyarwady Delta. In addition, we received news that the Burmese government had agreed to issue visas to 30 health workers from India, Bangladesh, Thailand, and China to support relief efforts. This is a welcome initial step, but concerns linger about the ability of these staff to travel freely in order to do their job. We also continue to urge full access to experts regardless of their country of origin.

The U.S. Pacific Command is prepared to provide relief supplies, including food, water, blankets, and medical supplies; equipment; and personnel to help in the relief efforts in Burma.

There are now a total of 11,000 service members and four Naval Ships in the region that could be used for aid to the Burmese people. Our Naval assets (USS ESSEX, USS Harpers Ferry, USS Mustin, and the USS Juneau) have arrived in the international waters off Burma. They are ready, but are still awaiting a request and approval for assistance from the Burmese authorities. Aboard these vessels are approximately 14 heavy and medium lift helicopters (CH–46, MH–60, SH–60) and more than 14,000 5-gallon plastic bladders filled with fresh water.

In addition, there are approximately a dozen cargo aircraft (C–130 and C–17), along with about a dozen heavy and medium lift helicopters (CH/MH–53 and CH–46) at Utapao Air Base in Thailand, that are available for the relief effort. The U.S. military is also available to transport other countries' relief supplies into Burma.

Our reports indicate that while supplies are slowly getting out to affected areas, at best 20 percent of the 2.4 million people who are affected had received any aid as of May 19. Only 250,000 of those 2.4 million people had received food aid through the World Food Program. Our own 20 relief flights of USAID commodities completed between May 12 and 19 will serve more than 106,000 people with merely a fraction of the supplies they need. Furthermore, health experts predict that if medical staffs do not have full access to affected populations, major outbreaks of disease could develop in the coming weeks, making this phase only the beginning of a much graver emergency.

Burma exemplifies the kind of situation where immediate access to food is critical and where having the authority to use up to 25 percent of the P.L. 480 Title II appropriation for local and regional procurement in P.L. 480 Title II could significantly improve the U.S. ability to respond effectively to this disaster.

HUMANITARIAN NEEDS

I have touched on some of the needs, but I would like to explain in further detail the extent of the aid that is required to help those affected by the cyclone. The U.N. Office for the Coordination of Humanitarian Affairs reports that critical needs include water and sanitation, food, shelter, and medical care.

Our leading concerns continue to be the provision of safe drinking water and the prevention and treatment of diseases associated with poor water and sanitation conditions. Aid agencies are extremely concerned about insufficient access to safe drinking water; a lack of water containers; and issues with sanitation and hygiene.

The U.N. World Health Organization (WHO) has reported that an increase in the area in which mosquitoes and other insects can breed is expected to result in higher than usual numbers of people becoming sick with malaria and dengue fever within weeks. Even during an average rainy season in the delta, incidence of these illnesses is high. If medical staff cannot get to those in need to provide preventative and curative care, the outcome may be dire. WHO and other agencies are pre-positioning drugs and rapid test kits, as well as distributing insecticide-treated mosquito nets where possible.

Urgent relief needs include safe drinking water, shelter materials, water purification tablets, cooking sets, mosquito nets, food, medicine, and sanitation facilities. USAID, as well as other donors and relief agencies, continue to airlift emergency relief commodities into Rangoon, but our efforts are not yet enough to begin to meet the needs.

Food security is also a major concern for relief organizations. Even in good times, the nutritional status of those living in the delta region is poor. The U.N. notes that in cyclone-affected areas more than 30 percent of children under five years old are chronically malnourished. This condition results from a combination of factors, including insufficient nutritious food, poor health care access, inadequate water and sanitation facilities, poor maternal and child care, and few livelihoods opportunities. As of May 18, the World Food Program had dispatched more than 1,240 metric tons of rice, high-energy biscuits, and beans to Burma's cyclone-affected areas. However, WFP estimates that only 250,000 people have actually received food. WFP estimates that it will need to move 390 tons of food every day if it is to reach the 750,000 people it is targeting over the next 30 days.

The cyclone hit at the beginning of the rainy season in Burma, which lasts from May until October. This means that heavy rains and harsh weather will continue to exacerbate conditions and hamper efforts to help the Burmese people begin to rebuild their lives and livelihoods, and plant their crops for next year's food supply.

USAID RESPONSE

The USAID response to the crisis in Burma has three components: technical assistance; relief supplies; and funding to the United Nations and our non-governmental organization partners on the ground working with those in need.

The USAID Disaster Assistance Response Team, which we refer to as a DART, remains in Bangkok, Thailand, where they are coordinating with DOD and U.N. and NGO partners. We have repeatedly sought Burmese government approval to deploy our team to Burma to support relief efforts; however, the Burmese government has not approved this request.

Our DART includes experts with various specialties. With many years of disaster response experience, the technical assistance personnel on a DART are usually relied on by local officials and implementing partners as expert consultants in disaster settings worldwide. In a large-scale international disaster, the most important role a DART plays is that of field-level hands-on coordinators. A DART can help identify gaps and priorities and efficiently target assistance to those who need it most. In addition, the DART can oversee firsthand the proper management and accountability of American taxpayer dollars, a responsibility we take very seriously.

When a DART is not present during a large-scale emergency, we are not able to adequately assess and coordinate efforts with our international counterparts and local officials. We cannot talk to those who are affected, to see damage firsthand. It takes us longer to determine the best course of action and use of our resources. We have to work hard to learn of gaps in assistance through third parties—requiring us to try to de-conflict third-party reports—and use our years of expertise to make sense of what we are hearing. This is exactly what our team is doing in Bangkok now.

Our team is also coordinating the movement of all US supplies to Burma, since Bangkok is currently serving as a staging ground for the entire relief operation, thanks to the cooperation of the Royal Thai Government. The provision of these supplies is the second component of our assistance.

Between May 12 and 19, the Department of Defense completed 31 airlifts of emergency relief commodities, including 20 airlifts of USAID commodities such as USAID stocks of hygiene kits, insecticide-treated bed nets to protect against malaria, containers for drinking water, and plastic sheeting for shelter. These supplies will help more than 106,000 people.

USAID Administrator, Henrietta Fore; Admiral Timothy Keating, Commander of the U.S. Pacific Command; U.S. Ambassador to ASEAN Scot Marciel; and DART team leader Bill Berger accompanied the first airlift of relief supplies to Burma on May 12. Administrator Fore and Admiral Keating were met in Rangoon by a Burmese delegation that included the Commander in Chief of the Burmese Navy, the Burmese Deputy Foreign Minister, and the Burmese Deputy Minister of Social Welfare.

The Burmese officials expressed their thanks for the relief supplies, and Administrator Fore and Admiral Keating talked about what further assistance the U.S. was prepared to offer. Both delegations agreed that water and food were top priorities. Our military forces have remained poised to support this effort for more than a week.

The supplies were handed over to the Burmese authorities who had helicopters assets capable of moving the much-needed supplies to the victims in the delta region. While on the ground in Rangoon, Administrator Fore and Admiral Keating witnessed Burmese helicopters, loaded with U.S. Government relief supplies, take off for the delta region. U.S. Embassy staff members have been monitoring the offloading of U.S. relief supplies on subsequent flights, and as of May 15 had not found any verification of the rumors that U.S. aid had been diverted. Others on the ground also have reported that they have not witnessed our supplies being sold on local markets or diverted. That said, U.S. helicopter capability could add significantly to the aid distribution effort.

On May 16, USAID began handing over supplies directly to non-governmental organizations, which have the ability to transport supplies into the affected areas and distribute them to beneficiaries. This shift will provide more accountability to ensure that U.S. supplies are getting to the Burmese people.

As I mentioned earlier, the provision of these supplies is the second component of our assistance. The third component is the funding we provide to NGOs and U.N. organizations that are working in the affected areas.

As of May 19, the U.S. Agency for International Development had provided $17.5 million in aid to Burma. Through the Office of U.S. Foreign Disaster Assistance, USAID provided an initial $250,000 on May 5. Those funds went to UNICEF, the U.N. World Food Program, and the Office of the U.N. High Commissioner for Refugees for immediate emergency relief assistance.

Another $1 million has been channeled through the American Red Cross, and more than $2 million channeled through World Vision, Save the Children, and Pact-NGO partners, who are currently working in the disaster region. OFDA is also providing $1 million to the World Food Program to support logistics services for humanitarian organizations. USAID's Office of Food for Peace is providing $12 million of P.L. 480 Title II food aid to be provided through the World Food Program.

When combined with the cost of relief supplies and funds to the Department of Defense for logistics, this brings total U.S. Government humanitarian assistance made available as of May 19 to more than $30 million.

THE WAY AHEAD

This is a disaster of immense proportions with serious consequences for Burma. Historical experience shows that even the best-prepared country facing a disaster of this magnitude requires international assistance.

USAID is working to fund NGOs and U.N. agencies that are currently operating in the affected areas. We will rely on these trusted partners to help us understand the humanitarian needs and provide appropriate aid to those who need it. We're also continuing efforts to persuade the Burmese government to allow in more experts and aid workers. We are closely monitoring reports that would indicate deteriorations in the health situation, and we are working to ensure relief supplies and services reach those who need them.

Above all, I want to emphasize that USAID still remains ready to help. We continue to stand by the people of Burma during this difficult time. Thank you, Mr. Chairman, for taking time to discuss this important issue. I would be happy to answer any questions you may have.

Mr. FALEOMAVAEGA. I want to thank both gentlemen for your insightful statements. Before doing so, we have another distinguished member of the committee also joining us this morning, my good friend, the gentleman from Arizona, Mr. Flake.

Did you have an opening statement that you wanted to——

Mr. FLAKE. No.

Mr. FALEOMAVAEGA. All right.

Before asking my questions—I am going to start the line of questions—I will ask my good friend from California, Mr. Rohrabacher, for his questions.

Mr. ROHRABACHER. Thank you very much.

In terms of the actual suffering that is going on in Burma, to what degree are we seeing that this is a natural calamity and suffering will naturally come after such things, and to what degree is that suffering the result of Government corruption and repression?

Mr. MARCIEL. Mr. Congressman, I guess I would answer it in three ways.

First, there has, as you know, been a long history, unfortunately, of repression and bad policies that have created a lot of suffering, both direct in terms of human-rights violations and indirect in terms of bad policies that have made life difficult for the average Burmese.

The cyclone is obviously a huge natural disaster. I am not an expert on such things. But clearly it would have caused significant damage and casualties in almost any situation.

I think the critical point and the point where we would be most critical of the authorities in Burma is their inexplicable failure to, A, make relief a priority and, B, to allow in the international help that they need.

So it is hard to quantify, but certainly every day that goes by and more people suffer, increasingly the blame falls on the Government.

Mr. ROHRABACHER. Could you compare this to what happened in Thailand just a few years ago when there was a tsunami that came in and, of course, affected thousands upon thousands of Thai people? Could you compare how the Governments reacted differently to the crisis at hand?

Mr. GOTTLIEB. If I might, having worked on that tsunami and also the Pakistan earthquake and many others, what I would say is there is really not much comparison between the two. After the tsunami, we had almost immediate access. The only place in which we were stymied for a little bit was in Aceh, and that was simply because there was a conflict going on there. We actually had people on the ground there within 48 hours.

I don't think there is really much of a comparison. The Thais were very open with us, not just in terms of providing assistance to them, but in terms of making U-Tapao Air Base available for us to fly into Indonesia. So we can only commend the Thais in the way that they have helped during the tsunami and during this crisis as well.

To go back to the question that you had for Mr. Marciel, I think one of the things that is important to remember is that in the Delta area there is about 30 percent malnutrition among kids to start with. So that alone gives you, sort of, the starting point at which we are beginning our work.

So children are already in a vulnerable state, and the fact that we are able to supply maybe 20 percent of needs at this juncture and probably very few of the special needs for kids, I think just increases the vulnerability of those children to disease.

Mr. ROHRABACHER. So when all is said and done, will we actually be looking at a situation where, yes, Mother Nature has caused the death of tens of thousands of people but an equal contributor to the crisis and the death of thousands of other people can be attributed directly to the actions or inaction of their Government?

Mr. GOTTLIEB. If I might make a comparison, in the late 1990s there was a crisis in—as you know, there was a crisis for many years in Sudan. There was a serious drought in southern Sudan, and about .25 million people died, not because there was a drought but because we had no access.

Mr. ROHRABACHER. Mr. Chairman, I would suggest that the Burmese regime is committing an act, a criminal act that should be recognized by the way they have approached those of us who would help their own people in this time of crisis.

It is a horrible situation. We saw similar natural disasters in Indonesia, in Thailand and other places, where the Governments actually worked with the decent people of the world to try to alleviate the suffering caused by the natural disaster. Here we have the Government going in the opposite direction.

This is criminal behavior, and it is time for a new wind to blow across Burma and to wipe away and blow away this regime that has suppressed their people for far too long. The Burmese people deserve so much more. And it is just very sad. They are very good people; I have been there. And let us hope that this terrible disaster brings about the change in their society, as it has in other dictatorships, like I mentioned with Samosa.

Thank you very much.

Mr. FALEOMAVAEGA. I thank the gentleman from California.

As a courtesy to my good friend, the distinguish ranking member of the subcommittee, I am going to have him offer his opening statement, whether he wants to do it for the record or he may want to expound upon it before I give my good friend, Ms. Watson, a chance to ask questions.

Mr. MANZULLO. Thank you, Mr. Chairman. I just want to take a few seconds.

A couple weeks ago, Congressman Joe Crowley from New York and I were at the White House at a signing ceremony with Senator McConnell and Senator Feinstein, awarding the Congressional Gold Medal to Aung San Suu Kyi. The President and the first lady actually took about 15 minutes after that signing—and their hearts are really broken, if I could use those terms, over the fact that American aid stands so close and so willing to help. This has been a special burden on the part of the first lady, who very seldom has spoken out on foreign affairs issues.

But I just want to bring that up. I know that perhaps our Burmese friends are listening, as they monitor what goes on, the fact that we are very concerned about the plight of the people in Burma. We commend you for your efforts in doing everything you can to help out.

Thank you, Mr. Chairman. I would like to submit the rest of the statement for the record.

Mr. FALEOMAVAEGA. Without objection.

[The prepared statement of Mr. Manzullo follows:]

PREPARED STATEMENT OF THE HONORABLE DONALD A. MANZULLO, A REPRESENTATIVE IN CONGRESS FROM THE STATE OF ILLINOIS

Mr. Chairman, thank you for calling this important hearing regarding the devastation of Cyclone Nargis. I am deeply saddened not only by the tremendous loss of lives, which the Red Cross estimates at nearly 130,000, but by the extreme short sightedness of the Burmese government in not accepting international aid. Without a well coordinated multinational relief effort, I fear that the death toll will climb significantly higher.

When this Subcommittee met last October, we focused on the military junta's violent efforts to extinguish the Burmese people's hope for democratic change. At that time the disturbing images from Burma were of military troops shooting at unarmed Buddhist monks taking part in what was later called the Saffron Revolution. Today the images from Burma are once again of anguish and pain, but this time the lack of compassion by the junta for its own people is beyond belief. Together these two instances show that it is the people of Burma that suffer most as a result of the junta's continued rule. There simply is no end in sight.

Thus, I commend the Administration for reaching out to Burmese leaders to encourage their acceptance of American and other foreign assistance. I understand that Admiral Timothy Keating, Commander of U.S. Pacific Command, made many attempts to allay the junta's fears of any hostile American intentions associated with American assistance. However, despite such efforts, the regime still refuses to allow any meaningful relief to enter into the country. With regard to the miniscule humanitarian supplies that have been delivered, I am extremely concerned by reports from the media that the junta was confiscating the goods and even selling it in the black market.

Mr. Chairman, I want to join the growing chorus of voices from all over the world calling on Burma's rulers to accept foreign assistance. Without help, death from disease, exposure, and hunger is sure to increase. We must keep in mind that Burma is just at the beginning of its monsoon season, from May to October, when heavy rains bring about seasonal flooding. If we don't help the poor Burmese people now we will watch hopelessly as more die from easily preventable deaths.

The U.S. has significant experience in providing disaster relief in Southeast Asia and the Indian Ocean as a result of our participation in the aftermath of the December 2004 tsunamis. Even in areas such as Aceh, Indonesia, where there was an ongoing separatist movement, America's intervention focused only on providing humanitarian assistance. Thus, my message to the junta is simple: open your doors now!

Two weeks ago, I stood in the White House Oval Office with my good friend Representative Joe Crowley of New York, Senator Diane Feinstein of California, and

Senate Minority Leader Mitch McConnell of Kentucky to watch the President sign into law H.R. 4286, a bill that I coauthored to award the Congressional Gold Medal to Aung San Suu Kyi. It was two weeks ago that Cyclone Nargis struck with all its might, and I am heartbroken to say that Burma still remains closed.

Thank you Mr. Chairman. I look forward to hearing from our distinguished witnesses.

Mr. FALEOMAVAEGA. The gentlelady from California for her questions.

Ms. WATSON. We have been following this disaster through the news, newspapers and media. One day you hear that the military government is not allowing any foreign workers in. Then, in the last few days, we heard that they allowed a few.

Can you clarify what is taking them so long to get visas? Why do you think they were resisting outside help? And what is the impact—and, you know, we know what we see from the news, but I understand there are villages that haven't even been accessed yet. So we don't know what the toll would be, and we don't know how many people are still alive.

So can you, kind of, give us some clarity on what to believe this Government is doing for its own people?

Mr. MARCIEL. I will try.

Ms. WATSON. The best you can.

Mr. MARCIEL. The fact is, it is not a clear situation, because we don't have a good picture of the situation overall in the country, or certainly what the Government leaders think. And because of the lack of access, we certainly don't have a good picture on the situation in the Delta, where the cyclone did most of its damage.

I would make a couple of points. This regime is extremely xenophobic, nationalistic, isolated—pick your term. The lack or the failure of people to get visas was apparently a decision by the Government not to let them in or, at a minimum, a failure to make a decision to allow them in.

I have been trying for a few years to figure out why they do what they do, and I don't think I have a great answer for you, other than they are more concerned about staying in power than they are about helping their people. That is clear from their policies overall. As I said earlier, it is appalling.

My colleague may have more information on the impact.

Ms. WATSON. Please.

Mr. GOTTLIEB. I think the only thing I can add to that is that, from what our team out there has heard, that the Government claims that it has the ability to distribute the aid. It only needs the things, not the people, to do that. That is not the observation that we have and certainly not what we necessarily hear from our colleagues who are in Burma. So, for us, it is a very poor decision that impacts on the lives of those that are affected by the cyclone.

Ms. WATSON. I notice that we have, from the chart, contributed $17.5 million. This is USAID Disaster Assistance Response Team and 13 airlifts.

Are those resources actually getting down to the people, or is the overhead of delivering taking up much of that? How effective has that amount of aid been?

Mr. GOTTLIEB. To date, we have managed to do—with DoD there has been, actually, 36 C–130 flights going to Rangoon. As Mr. Marciel noted in his testimony, the first flights were handed over directly to the Government. But since the 16th, we are handing it over directly to our NGO partners. These are partners that, of course, are familiar to you, as I mentioned, partners like UNICEF and others. And they are moving that food, they are moving that directly into the areas where they have been working.

I think what is important to understand about Burma is that, even before the cyclone, there were, roughly, for us, from what we can tell, about eight NGO partners that we are very familiar with, both American and European. We have a lot of confidence in those groups. Some are quite large like World Vision, which has 500 local staff.

Ms. WATSON. Were they on the ground during the cyclone?

Mr. GOTTLIEB. Yes.

So, first of all, we have confidence that they wouldn't take it if they didn't think they could distribute it. I think that is number one.

Number two, at least for our U.N. partners, we are aware that they are expanding their staff. World Food Programme now has 200 local staff; they have 1,400 internationals, although those internationals are largely confined to Rangoon.

But we do feel that the aid that we have now is making its way out to the field, yes.

Ms. WATSON. Thank you.

Thank you, Mr. Chairman.

Mr. FALEOMAVAEGA. The gentleman from Arizona, Mr. Flake.

Mr. FLAKE. I apologize for not hearing the testimony. I have tried to read through as much of it as I could.

You mentioned you are working with NGO partners here. Are there any U.S.-based NGOs that are approaching the government there directly and having more success, or are the Burmese treating United States-based NGOs just like they do our Government?

Mr. GOTTLIEB. To their credit, I believe Samaritan's Purse did manage to get permission to fly in—I believe they had one plane-load go in. I think it was quite large. It might have been a jumbo jet, I am not sure. No, it wasn't a jumbo jet, but it was quite large. But I don't know of any other American groups that have had that direct access yet.

Mr. FLAKE. Okay.

Mr. Marciel, you mentioned that you have tried to monitor the supplies that we did ship in to see if some were being sold or what-not, and we have not found indication that they are.

Are we assuming, then, that they are hitting the target for the most part? That seems to be the testimony

Mr. MARCIEL. It has been a big question on our minds.

When we initially got in with C–130 flights, we could not get approval from the Burmese authorities to deliver it directly to NGOs, which is our normal practice or USAID's normal practice. We made a decision that, because of the urgency of the situation and the need to try to build a little bit of confidence so that we could get more aid in, that we would turn over some initial supplies, mostly water and mosquito nets, to the Government. And they actually took us, the day I went in with Admiral Keating and AID Director Fore, they actually took us, the Burmese, to the part of the airport where they loaded our supplies on the helicopters, and the helicopters took off, we were told, to the affected area.

What our Embassy has been doing is reaching out to U.N. and others on the ground there to try indirectly to get a sense of whether materials are reaching the people. And they have also been scouring the markets to see if our materials turned up in the marketplace, which, of course, would be evidence that they had been diverted. They have not found such supplies in the markets. It is a good sign. It is not, you know, perfect. It is not a guarantee.

Mr. FLAKE. It is possible that they could be simply hoarding them.

Mr. MARCIEL. It is possible. That is why we are pleased that, since May 16th, the USAID materials we have been able to turn over to international NGOs. That greatly enhances our confidence.

Mr. FLAKE. With regard to the referendum, it was mentioned that it was postponed in the affected area until May 24th. Do we expect that to happen at that time, or will it be postponed further?

Mr. MARCIEL. Everything we have heard suggests that they do plan to go ahead with it on the 24th. How on Earth you would do it in an area that's been so devastated is beyond me. But, then again, the whole thing is a sham.

Mr. FLAKE. Right. Thank you.

Mr. FALEOMAVAEGA. The gentleman from New York, Mr. Crowley.

Mr. CROWLEY. Firstly, I failed to have mentioned, thank you for allowing me to participate, although I am not a member of the subcommittee, I am a member of the full committee. I always want to thank the chairman for allowing me to participate.

I agree with my good friend from California, Mr. Rohrabacher, on many of the comments that he made and addressed to the panel before and to the subcommittee. It is hard for me to even know where to begin. But some things that stirred my memory a bit were, again going back to the tsunami, I had the opportunity to visit Sri Lanka after the tsunami just about a week and a half after it hit, and I saw many, many assets on the ground, including United States, and in fact, they allowed an Indian ship, a military naval ship, to enter into Sri Lanka within days of the event itself. I think India may have been one of the only countries that actually didn't accept foreign assistance or didn't ask for it because I think they had the assets, the ability to address the issues themselves within the confines of their own country. But think at the time that Sri Lanka, albeit it was the peace process going on, the Tamil Tigers suspended activities to allow for assets to be delivered to that region. And again, in a matter of days, in Aceh, again, given the situation, the revolutionary situation within those areas, that assistance was focused on getting that to the people in need.

And then you contrast that with what has happened in Burma, including what really hasn't been touched on all that much has been this referendum that took place, when people were suffering during this time, when many of the people who were being asked to vote on the referendum on the Constitution, what state of mind could they have been in at that point in time. It is just mind-boggling to us that somehow the referendum would be viewed in a legitimate fashion.

Anyway, regardless of whether it was legitimately conducted or not, I also remember the refugee camps, for lack of a better word, or survivor camps in Sri Lanka, and the stories, you know sometimes we look at this and we just, it is so overwhelming to look at that, but then you kind of narrow it down to individuals as best you can. And the vulnerability, especially of women and children during these circumstances, and how they inevitably are abused, not necessarily by the government directly but by people who would take this opportunity to extract revenge or to covet someone they have always wanted to. And it is just the lack of attention by the government to protect the most vulnerable is probably the most disturbing aspect of any type of situation like this, especially when the state itself is failing to protect its own people. But Mr. Marciel if you could maybe just give us a thought, and maybe Mr. Gottlieb if you could expound upon it, what would have been the difference in your mind had U.S. assets been able to service the disaster area? What difference do you think we could have made had we gotten in right away?

Mr. MARCIEL. Congressman, perhaps I can offer an initial comment and then ask my colleague to say more?

Mr. CROWLEY. Sure.

Mr. MARCIEL. Admiral Keating, when we went on the very first C–130 flight, laid out very clearly to the Burmese

authorities what we were willing to offer in terms of helicopters, boats and other assets to carry the shipments, and was very forthcoming in terms of being flexible so that they would be able to keep an eye and monitor us if they wanted. So certainly we could have provided large amounts of assistance and gotten it to the affected area. Quantitatively I am not sure I can give you a number on how much—my colleague can ask—but certainly there is no question that not only the United States but other international, United Nations and others, could have gotten water and expertise into the area very quickly and alleviated some of the suffering. I will leave it to my colleague to see if he has more details.

Mr. GOTTLIEB. Thank you. Congressman, if I can compare this to the tsunami. The Essex is just one of the ships that is now offshore of Burma and was one of the ships that we used in Aceh. That carries approximately 23 helicopters. I believe 20 of them are actually usable. Some of them are quite large. They would be, in a situation like this, extremely helpful. First of all, one of the things that you try to do in a disaster like this is not have people move from their homes. You try to keep them where they are. People can begin to recover more quickly if they are in place. They can salvage what is left from their homes to begin to rebuild. But if you cannot bring assets to them, they will leave to go where those assets are. And that is what is happening now.

If we had helicopters, whether we could helicopter out of Burma, out of Thailand or off Essex, we could be moving supplies in there quite rapidly, and we could move it to various points quite easily. The importance of having people on the ground is to then plan for where those helicopters need to go. The importance of the assessments is that you know who needs what. And so we would develop a logistics system out of probably Rangoon, which would then determine what goes on each helicopter during the day. And that is what we had in Aceh. We had what we called a logistics cluster, a logistics cell inbound to Aceh that sent out orders to all the helicopters as to what they lifted and where they did. So what you would be doing is bringing the right thing to people, and you would be restarting the whole process of helping them restart the process of their lives.

Right now, as I think the Congresswoman noted, now we are getting close to the time to plant. We would be looking at making sure that there was seed available. That may become part of a package that you would bring in to people to make sure that they have what they need to plant.

Mr. CROWLEY. Mr. Chairman, can I ask just one simple question? I know I have been indulged here already, but could we have saved lives had we been given access earlier?

Mr. GOTTLIEB. You know, I don't have a report of anybody dying. But I think the important thing to note is that in a place where 30 percent of the children are already malnourished, what you want to provide for children is high energy foods to help maintain their health. So I think what is important here is that you access people quickly. You know, as I said, I can't confirm one death or many deaths. But I will say that we know that when kids are malnourished, you have to intervene with the right thing at the right time.

Mr. CROWLEY. Thank you.

Thank you, Mr. Chairman.

Mr. FALEOMAVAEGA. I thank the gentleman from New York.

I have about a hundred questions for you. I wanted to give my colleagues every opportunity to raise questions. We have a saying in the islands, gentlemen. [Speaking in native language.]

Mr. FALEOMAVAEGA. Which means, the leaves of the coconut tree just doesn't sway by itself. There is a reason for it. There is a cause.

And I believe I have a more fundamental question I wanted to raise with you both. The very fact of the existence of the current military regime, if you will, and correct me, Mr. Secretary, and my simplistic understanding of the problem of the history of Burma, even the British could not colonize these people. In fact, the word Burma came from the British Government at the time when they tried to colonize this country. My understanding is that because Burma is composed of about seven or eight basic groupings within the government, that no government was able to ever unify these people unless and until or I suppose when they had a good military to take control of these different factions. Am I correct in this, Mr. Secretary, or am I wrong? Am I way off the record on this?

Mr. MARCIEL. Mr. Chairman, my understanding is that the British when they were the colonial power did have pretty extensive control. But you are absolutely right that there are a lot of different ethnic groups and that there has been, certainly since independence, separatist insurgencies carried out by a number of those groups.

Mr. FALEOMAVAEGA. I remember in my youth, there is one word that stood out in my going to high school, was the name U Thant, probably one of the greatest Secretary-Generals of the United Nations as ever, who happens to be from Burma. What took place in

1962 and since that time, the military takeover of that country, what I am trying to get here, Mr. Secretary, it seems that the strongest condemnation in the current regime has come not only from the United States but from its allies. Am I wrong on this?

Mr. MARCIEL. No, you are right.

Mr. FALEOMAVAEGA. What about our fellow Asian countries, like members of the ASEAN that Burma is a member of? Have they also been very critical of the way we have gone about in criticizing the current military regime?

Mr. MARCIEL. In the past, they have been somewhat critical. I would say less so over the last year. You will recall that in, I believe, September or October of last year when the military government cracked down against peaceful demonstrators, ASEAN spoke out very strongly, frankly, in unprecedented terms condemning the regime's actions.

And I know from having to spend much of the last 3 or 4 weeks in the ASEAN region, I don't want to put words in their mouths, but my sense was that many of the ASEANs are also appalled at the Government of Burma's failure to respond to this crisis.

Mr. FALEOMAVAEGA. And currently, the trade relationships between Burma, more so with the member ASEAN countries like Thailand, and probably no other country more influential on the leaders of Burma at this point in time is the People's Republic of China. My question here is, why isn't our Government pursuing this issue before the Security Council of the United Nations on the issue of humanitarian assistance alone? Have we ever taken the initiative before the United Nations Security Council that this issue is such an international—this not just Burma alone. We are talking about cholera. We are talking about disease. We are talking about millions of people, even to the point that this is genocide. And I wanted to ask you, has our Government taken the initiative to bring this before the Security Council, which is the, I suppose if something comes out with the Security Council, the United Nations will listen, especially because of the five permanent members that have all the muscle and the resources to carry through these things?

Mr. MARCIEL. Mr. Chairman, we have consistently argued that the overall situation in Burma warrants Security Council attention, and we have brought it to the Security Council many times. Since the cyclone, I think others—I think it was the French who raised this in the Security Council. We supported that discussion. Some in the Security Council argued that it was, while the disaster was a disaster and warranted international attention and U.N. attention, it was not appropriate for the Security Council. We continue to believe the situation in Burma warrants Security Council attention, yes.

Mr. FALEOMAVAEGA. But didn't France also throw a curve ball in this whole idea of assistance and say, well, we will do it only if it meets our conditions and our requirements in giving assistance? Wasn't that one of the problems also?

Here is my concern. If I were a member of the military regime, after being condemned that it is the worst, why would I want you to come help me when you make me look so small? In the eyes of the nuances of the cultures that are in that region, shooting words and attacking in verbals is almost like cutting my right arm or my leg off. This is like: You have dishonored me so much, why should I accept your offer of assistance when you made me look so small before the international community?

Mr. MARCIEL. Mr. Chairman, I think the only answer I can give is that when this cyclone hit, the United States, which as you pointed out has been probably the harshest critic of this regime, from the President on down, we said we had our political differences, but this is a humanitarian disaster and we are setting aside for the moment those political differences to focus on saving lives of upwards of 2 million people. The regime may or may not like us and certainly didn't like the criticism. But, frankly, the reason for them to accept international help is because otherwise tens of thousands, perhaps hundreds of thousands of their own people may die. And it seems to me that it is incumbent upon any government to take steps necessary to protect its own people.

Mr. FALEOMAVAEGA. I have the highest respect for our First Lady and the fact that she was there, the initiator, the leader, as far as the United States, I guess representing the State Department and even the President when she first announced our willingness to give assistance but at the same time right along with it condemning the military regime. And you know, we all love Aung San Suu Kyi. I mean, everybody is very, very appreciative and admire her for her leadership in all this. But it kind of gives it at a different mix there when you say we are going to give humanitarian assistance, but we continue condemning, which again this is what democracy is all about here in America. But in that part of the world, as far as they are concerned, you have cut my right arm off, so why should I accept your assistance. I would rather die than to receive assistance from you because you have dishonored me.

Ms. WATSON. Mr. Chairman, would you yield for a second?

Mr. FALEOMAVAEGA. I will be glad to yield to my friend from California.

Ms. WATSON. You know, I am sitting here listening to all of us, and it goes to our foreign policy. And you know, over the couple of weeks that this has been number one in the news you sit there and you watch and you say, Why would they not want the humanitarian help that we are offering? Well, it goes to our policy. On the one hand, we will give it and we are pulling it back by condemning you. So I think it is time now that a group of members take a trip and talk to whatever leadership we can in China and in Myanmar about how best to help when help is needed and take away the politics of it and get on the track of humanitarian assistance. Our country has to take the lead in that. While we are sending resources to another part of the world that is going down into a gofer hole, our resources really ought to help people. I think that speaks louder. But we have to figure out how we can separate from the politics of all of this. I think their resistance is against the Western power that they feel are imperialistic.

We truly want to help people. We want to help the children. We want to help the women. We want to help the people, and that has to get across. I just had to enter that into the conversation.

Mr. FALEOMAVAEGA. I thank the gentlelady for her comment, and I want to ask Secretary Marciel, What do you recommend that Congress should do?

Mr. MARCIEL. Well, Mr. Chairman, I would like to comment on this because I think this goes to the heart of the matter. We have been very clear over the last few weeks that, despite our political differences with the regime in Burma, we are interested and very sincere about offering humanitarian assistance. And I think we have demonstrated that sincerity through our actions.

But it is more than that. This is not just a matter of the regime saying no to help from the United States. There have been a series of envoys from the United Nations, from many different countries, who have gone and urged the authorities to accept international help, not just from the United States, from the U.N. humanitarian agencies, from NGOs. This is a regime that is not just saying no to the United States because of political differences. It is saying no for the most part to help from the entire world; from the region, from the United Nations, from everyone. So if I could, I would urge that we not look at this as a United States/Burma issue. It is much bigger than this. This is an issue of the Burmese regime failing to take steps to protect its own people.

Mr. FALEOMAVAEGA. Are members of the ASEAN association there helping Burma right now?

Mr. MARCIEL. They are trying. The ASEAN foreign ministers called an emergency meeting which took place yesterday in Singapore. Out of that came an agreement that ASEAN countries would be able to lead an international relief effort. The ASEAN Secretary-General, Mr. Surin Pitsuwan, went into Burma today to try to work out the details of that. We actually spoke with Dr. Surin last night just before he went in. The details are still unclear. They have to work this out. We hope that this is going to lead to a significant increase in assistance, including via ASEAN, and we welcome ASEAN's efforts. But at this point, we don't know yet whether the Burmese regime is actually going to allow them to bring in the experts, the equipment and other international help.

Mr. FALEOMAVAEGA. What I am trying to convey here, Mr. Secretary, is that Burma is a member in full standing with the ASEAN, this regional organization there in Southeast Asia. And they do active trade with Thailand, even with Singapore, I am told, and of course China and other countries despite the fact that this is a military regime that we are looking at. So how is that, what is that perception? I mean, it is a military regime, but yet the fellow ASEAN countries are doing business with them just like nothing is happening.

Mr. MARCIEL. Again, I don't want to speak for ASEAN. I can give you my sense of what is happening.

Mr. FALEOMAVAEGA. Please.

Mr. MARCIEL. The ASEAN members have argued for a long time that the way to try to encourage positive change in Burma is to work with the regime, try to convince them that they can move in a different direction. So that has been their position. They don't believe that sanctions work, and they believe that continuing to work with these people is the best way to try to move them in a more positive direction. And more recently, of course, in the aftermath of the cyclone, they, like the rest of the international community, have been using whatever influence they have to try to urge the generals to allow greater access.

Mr. FALEOMAVAEGA. I was asking one of the foreign ministers there representing ASEAN, I expressed my commendation to them for getting organized. And he said, you know, it took us 10 years just to say hello. I mean, just to show how complicated, not as simple and as easy as we perceive it to be; he said it took us 10 years just to get together, just to say hello to each other among the ASEAN countries. So I can imagine when it comes to trade and other issues that deal with this regional organization, that it is a very complicated one. And I am just curious—actually, my good friend from New York probably has a question.

Mr. CROWLEY. Mr. Chairman, if you would yield for a moment because I just want to go back to what you had initially talked about when you began the questioning in regards to how we are perceived by the folks, the people within Burma, the junta, and I guess a lack of tolerance, or we have great intolerance towards, from here, for the Western perspective on what they are doing to their own people. And I think not being sensitive, I guess you can say, I want to point out to the folks, I think the chairman has a great deal more sensitivity, I think, being the closest geographically in his representation and has a great understanding, I think, of cultural as well as sensitivities, and I think that is why he adds such a great deal to our membership here in the House.

But what I would just suggest is, you know, timing is an interesting thing. We have been working on the passage of a gold medal for Aung San Suu Kyi for some time and passed the House in December of last year and passed the Senate sometime in April, I believe, of this year, and the timing of the signing of that legislation by the Speaker to forward the bill to the President happened to take place prior to the cyclone hitting and in many respects overshadowed what was a timing issue. The President had to sign the bill within I think a 10-day period. And in many respects, some of us were working on it for some time, and it was frustrating because it took a great deal of the oomph away from what we were trying to accomplish and at the same time noting that the junta itself did not postpone, total political timing, did not postpone its own referendum. I don't pretend to have an expertise on the cultural aspects of the Burmese junta or the people, but I can express I think a frustration that at least many within the Burmese community here in the States have towards, a resentment toward this junta that precedes the cyclone and will have after the cyclone may be long forgotten. So, with that, I just want to yield back, Mr. Chairman. Thank you.

Mr. FALEOMAVAEGA. I want to say to my good friend from New York, this one of the strengths of our great democracy in our country, is this diversity. And my good friend, I know every Irish-American here is very proud of the fact that we hosted recently the Prime Minister of Ireland in a Joint Session of Congress, and the fact that for years the Northern Ireland situation that has been the civil war, if you will. Even Senator Kennedy and all of our Irish-American leaders were trying very hard to see the final resolution to the problems.

The situation that we also have with North Korea, that the People's Republic of China was very instrumental in bringing North Korea to the table. And these are some of the nuances that I think, for myself, yes, I am a very, very proud American, but my roots are from the Asia-Pacific Region. And maybe this is a reason, when I say the leaves of the coconut tree just don't move by itself, an essential understanding and appreciating some of the nuances that happen there, why is it that we do things maybe in a different way? And by this miscommunication, perhaps, or misunderstanding is carried on to the point where there is no communication and no resolution to the problems. And that is a reason why I raise the issue. If we have some understanding or appreciation of the root causes, why we are where we are now, and the fear that perhaps the military leaders of this country have that there will be anarchy without them. At least that is their position, that without the military, Burma will dissipate, seven or eight different factions will be killing each other off. That is their reason for existence, I suppose.

But in understanding that, why Aung San Suu Kyi and her party was not given an opportunity to rule, not that is—now, here again, that is a different situation, too, that we find ourselves in, one I do understand a little more about what happened. I am told that a member of the Armed Foreign Service recently attended this referendum in one of the polling stations and that the military leaders made use of his or her presence. And if I see, America supports the proposed Constitution. Can you care to comment on this Mr. Secretary? There was a Foreign Service Officer, a member of the diplomatic corps, representing the United States, went over to, I guess to observe this referendum, and then, by doing so, the Burmese Government made a real good adage to say, hey, see, even America is here to witness the way that we have conducted the referendum, and that is why it passed by 90 some percent. Can you comment if there really was a Foreign Service Officer that did this?

Mr. MARCIEL. I don't know the specific case. I do know that at the last minute, the Burmese regime did invite diplomats in Rangoon to send out a couple of people from each Embassy to observe. We did send out a couple of people just to see what was going on. I don't know the details of the specific case. It wouldn't surprise me if the regime used this for propaganda purposes. We do this around the world. When there are elections, we routinely send our Embassy people out to try to get a sense of what is going on on the ground, and so we did.

If I could, Mr. Chairman, one comment on your previous statement about the military's role. The military in Burma has argued for some time that it needs to be in power, the one force capable of running this diverse country. We, of course, would argue that isolating yourself from your people and refusing to honor the results of election is not the best way to unify. But I think also this cyclone demonstrates the fallacy of the military's argument. I can't imagine a worst and more negligent response to a natural disaster than what this supposedly well organized military has done. So I think it shows very clearly the weakness, the hollowness of this claim.

Mr. FALEOMAVAEGA. Are you familiar with the provisions of the proposed Constitution that is now being passed in the referendum? Are there any provisions there that are highly questionable in terms of the implementation of this newly accepted Constitution in the referendum? Are you familiar with any of the provisions of the proposed Constitution?

Mr. MARCIEL. I am familiar with some of them. What I would like to do, if I could, is give you a brief answer and then maybe get back to you with more details.

There are a number of issues or provisions in the Constitution that concerned us: One, that will reserve, I believe, at least 25 percent of the seats in Parliament for the military; second, that require the President of the country to have a military background; third, that would prohibit anyone who has or has had a foreign-born spouse from serving, for example ruling out Aung San Suu Kyi; fourth, a provision that would allow in effect the military to take over at any time that it thought there was an emergency. So there are a whole series of problematic provisions.

The other broader point here though was that the Constitution was drafted in secret by a group handpicked by the regime with no input from the democratic opposition, so the whole process was not legitimate in addition to the specific substance of the Constitution.

Mr. FALEOMAVAEGA. Without objection, I would like if you could submit a copy of that proposed Constitution, every provision that will be made. I want to make it a part of the record.

Was it a fact that because Aung San Suu Kyi married a British citizen a factor among the people there in Burma in terms of its political situation there? Did that have a bearing on some of the things that went on there that kind of made somewhat the issue divisive in that sense?

Mr. MARCIEL. Mr. Chairman, I don't have the sense that this was an issue that resonated with the people of Burma. The only real evidence for that, though, would be to go back to the 1990 election, the last time there was a real election in Burma, in which, although Aung San Suu Kyi

herself was not allowed to run, she led the party that won I believe about 80 percent of the seats. So the last, if you will, representative opinion survey in the form of an election showed overwhelming support for the party that she leads.

Mr. FALEOMAVAEGA. And the good part of her life, she was educated in England, was she not?

Mr. MARCIEL. Yes.

Mr. FALEOMAVAEGA. And then she returned married to a British citizen. You don't think that was a real factor there in the minds of the Burmese people in how she carried herself as a leader among the people?

Mr. MARCIEL. It doesn't seem to have reflected in any way, certainly in the election results or in the crowds that she drew when she was not under house arrest and was able to go out and talk to people.

Mr. FALEOMAVAEGA. Mr. Gottlieb, I didn't mean to disregard you, 25 years of humanitarian assistance. Do you think in any way possible that we can separate the politics from strictly on this one issue humanitarian assistance? Let us not talk about Aung San Suu Kyi, let us not talk about how terrible the military regime is, let us just talk about humanitarian assistance. Do you think that maybe that might give a sense of trust by the military leaders to allow foreign countries like, especially like our country, to give assistance, that we have the resources?

Mr. GOTTLIEB. Well, I think that the original flight that went in with the Secretary here and Admiral Keating and Administrator Fore, I think there was a message they tried to deliver. What we intended was to deliver humanitarian assistance based on need, not based on any other sort of precondition, strictly need. And I think we have tried to carry that message in other types of humanitarian assistance we have provided. For instance, if you will recall the Bam earthquake in Iran. Our DART team departed and arrived there within days of that earthquake. And I think that was a good example of how, at least in some instances, we can suspend our politics for a minute, both sides, for the best for the community. So I always, as a humanitarian, hope that is possible. I am not naive enough to believe it always works out that way.

Mr. FALEOMAVAEGA. I think sometimes in our art of diplomacy, words can be very cutting. I don't know if you watched the CNN documentary that Ms. Amanpour did on China when our philharmonic orchestra went to North Korea to give a concert. And in that documentary, a North Korean young lady was playing this instrument, and one of the things that she caught, at least in her mind, was when our President accused North Korea as an axis of evil power, she felt offended. And here is an innocent young lady from North Korea, a sense of pride, I suppose. It is almost like you calling somebody an SOB, excuse the expression, you feel hurt even though how bad the government may be, but individually words can be very cutting sometimes in how we express ourselves, especially as diplomats and as leaders of our own country. But do you think by giving a better sense of trust by the military regime that this eventually is coming about, especially from any pressures? I suppose, would you suggest that maybe China is probably the biggest influence of Burma to ease up on some of these issues that we are concerned about in giving assistance?

Mr. GOTTLIEB. Mr. Chairman, it is hard for me to comment on the broader aspects of this. I cover humanitarian assistance around the world. I haven't focused on one particular place, so I would defer to my colleague on the specific relationships. What I will say is that we were pleased when, after the first flight, we got more flights in, and then 4 days later after that one, we were at least able to turn over supplies directly to our NGO partners, which is for us a significant improvement. One, at least we saw a modicum of change in the regime stance. And two, doing so we feel allows us to feel better about where we think that assistance is going to go and the timeliness with which we know. We do have an instance from one of our partners citing the fact that they brought in assistance to Rangoon, and within the same day, it was down to the field and delivered. So we are, at least we are seeing a little bit of improvement.

Mr. FALEOMAVAEGA. I apologize to my colleagues. I think I have taken too much of my own time.

The gentleman from New York. Okay.

Diane, did you have more questions for our witnesses?

Ms. WATSON. Can you, based on what we have said here, tell us what you consider our next move to be that might bring about the kind of effect that we want?

Mr. MARCIEL. I don't see a dramatic move. I think what we have seen in the last several days are some signs of an increased willingness on the part of the regime to allow international relief in. We are in one of those situations where the opening has gone, if I can use the metaphor, from an inch to a foot. Probably that is even an exaggeration, maybe an inch to four inches, but it needs to be a yard. And so what we need to do, all of us, and when I say all of us, I mean not just the United States but the international community, the United Nations, is continue to talk to the authorities in Burma and try to convince them of the urgency of allowing in experts and equipment so that aid can get to the affected people. That is what the U.N. Secretary-General is flying to Burma today to talk about, the Secretary-General of ASEAN, and that is certainly our talks. We had a Marine three-star general in Rangoon today that met with regime officials, again reiterating all the ways we can be helpful. I think it is not a very good situation, there is no easy answer, but we think this is the best way forward, is to try to push as hard as we can to try to convince them to accelerate the speed with which they are beginning to open.

Mr. FALEOMAVAEGA. For the record, Mr. Secretary, we do have formal diplomatic relations with Burma; do we not?

Mr. MARCIEL. We do.

Mr. FALEOMAVAEGA. So that means any of us can go and visit using a visa?

Mr. MARCIEL. If you are able to get a visa, yes, Mr. Chairman.

Mr. FALEOMAVAEGA. I don't know if anybody here is representing the Burmese Embassy, but I have been trying to get a visa myself for the whole last year. And if you are hearing me about this, please report back to your Embassy. I have been wanting to get a visa to go to Burma, and you have not given me an opportunity to visit your country.

Mr. Secretary, Mr. Gottlieb, thank you so much for your patience. We have our next panel. You can stay with us. I know you have a million other things you have to do, but thank you both for coming this morning.

For our next panel we have with us a very distinguished gentleman here, our former colleague and Member of Congress, Congressman Tom Andrews. Good to see you, Tom.

And also the Prime Minister, Dr. Sein Win.

Before their presentations, I want to just introduce to my colleagues the distinguished gentlemen. Dr. Sein Win was the elected Prime Minister following the formation of the National Coalition Government of the Union of Burma in Manerplaw or Karen State on December 1990. He was elected representative from the Paukkaung Constituency in the Pegu Division in Burma during the 1990 elections. He is the son of U Ba Win, one of Burma's top political leaders and elder brother of General Aung San, the architect of Burma's independence and founder of the Burma army and first cousin to Daw Aung San Suu Kyi, a leader of Burma's democracy movement in 1991 and the recipient of the 1991 Nobel Peace Prize.

Dr. Sein Win earned his doctorate in mathematics from Hamburg University in Germany. He taught at Colombo University in Sri Lanka, as well as in Rangoon University in Burma. He became involved in politics when the military cracked down on the people involved in the pro-democracy uprising of 1988. He is the chairman of the Party for National Democracy, a party which Aung San Suu Kyi and U Tn OO as patrons was set up in 1988 as a back-up party if the military authorities decided to ban the National League For Democracy. Dr. Sein Win is serving his fourth term now as the Prime Minister of the National Coalition Government of the Union of Burma.

Congressman Tom Andrews is a former Member of this body representing the great State of Maine. Congressman Andrews currently is president of the New Economy Communications. It is a nonprofit organization that provides strategic planning and communication services to individuals and groups working on human and labor rights issues at home and throughout the world. His clients have included various different organizations, not only in this country, worked very closely also with former Secretary of State Madeleine Albright. Congressman Andrews also works on behalf of Aung San Suu Kyi, a leader of the National League of Democracy in Burma.

And, gentlemen, we are very, very happy and honored to have you both come and testify in our hearing on Burma this morning.

And I would like to extend the courtesy to Prime Minister Win. Would you begin with your testimony?

STATEMENT OF SEIN WIN, PH.D., PRIME MINISTER, NATIONAL COALITION GOVERNMENT OF THE UNION OF BURMA

Mr. WIN. Thank you, Mr. Chairman. Before I present my oral statement, I would like to submit my written statement and two news articles we wrote in the International Herald Tribune and Asian Wall Street Journal concerning the situation in Burma.

Mr. FALEOMAVAEGA. Without objection, your statement will be made part of the record and any added materials you want to submit as part of your statement will be made part of the record.

Mr. WIN. Thank you, Mr. Chairman.

Please allow me to express my sincere gratitude to the Subcommittee on Asia Pacific for organizing this hearing at this crucial juncture when my country, Burma, is facing a national calamity of an unprecedented nature in its modern history. The whole country and the people of Burma overseas are mourning for the millions of people affected by Cyclone Nargis. When the storm hit the densely populated area in lower Burma on 3rd May, the cyclone killed tens of thousands and ravaged houses, farms and infrastructures leaving a lasting imprint of destruction which will take months, if not years, for the millions of survivors to recover from.

Even though the death toll cannot be confirmed with certainty yet, the estimation so far is that over 100,000 people are dead or missing and nearly 2 million people will suffer from the effects of the cyclone.

Even under such circumstances, the Burmese military junta has done very little to help the people. Its main focus since the cyclone had been to hold a national referendum so that a Constitution it had written to legitimize military rule would pass. At the same time, it is also preventing the international community from entering the country with the fear that the rigid control that the military has imposed on the country will be undone by the presence of international relief experts. What is even worse is that reports are trickling out about the international aid being embezzled by the authorities. The Burmese generals' short-sighted policy has worsened the situation for the cyclone victims.

Mr. FALEOMAVAEGA. Mr. Prime Minister, I didn't mean to intervene, but just put your mike a little further away from you. Okay. Let us try it now.

Mr. WIN. An international NGO working inside Burma estimated that 30,000 children are starving and the under-5s living in the area of the Irrawaddy Delta, who were already acutely malnourished when the cyclone hit Burma, might be dying from the lack of food now.

The five regions struck by the cyclone are predominantly agriculture zones where agriculture families producing 65 percent of the country's rice are living. International agencies have time and again been warning that time is running out for us, and Burma will be subjected to the risk of famine if the agriculture system is not restored soon.

We understand that ASEAN countries have now formed a regional task force to distribute foreign aid and that medical teams from ASEAN will be working in Burma. What I want to emphasize here is the formation of the task force and admittance of ASEAN medical teams into Burma should not be an excuse for the Burmese general to delay assistance or prevent international experts from joining the relief efforts.

Lest people have forgotten, let me remind that when the last tsunami struck East Asia it was international experts that the ASEAN nation had to invite for the relief and rehabilitation operations. In other words, ASEAN alone cannot help our cyclone victims, and Burma needs international relief officials with the right expertise to cope with the challenges.

From our point of view, delivery of humanitarian assistance to Burma is already too late. The relief and rehabilitation program should start now without any delay, and we support any move which will make that happen, regardless of what the Burmese generals think. This is because we are talking about the lives that are at stake now and in the foreseeable future. It will not be the natural disaster but the xenophobic Burmese general who will be responsible for killing them this time. The situation is totally unacceptable, and it must not be allowed to continue.

We are therefore calling the United States and all donor nations to do everything they can to start massive relief operations immediately. The leading political party, the National League for Democracy, and elected representatives in Burma who are witnessing the current situation have also called for immediate relief for the cyclone victims.

As a citizen of Burma and an elected representative, I want to stress that pursuing a diplomatic option to convince an intransigent regime like the Burma generals is like waiting for people to die, and time is something that the people of Burma do not have. Please help Burma now.

Currently, the military junta is more occupied with its referendum and Constitution than the plight of the people. It has announced that its Constitution has been approved, and the process to legitimize military is already in the works. We foresee even greater political instability in the coming months as people are bound to be frustrated with the worsening socio-economic crisis caused by the effect of Cyclone Nargis and the regime's lack of regard for human suffering.

Given the fact that the referendum did not reflect the will of the people since it was methodically manipulated by the military, I strongly ask the United States to reject the outcome of referendum and the military junta's Constitution.

Based on the research on famine epidemic outbreak that has happened in the world before, Amartya Sen, a well-known economics and Noble Laureate constructed a thesis that population and the authoritative regimes are more prone to famine epidemic outbreaks after a natural disaster than those under democratic governments. Burma after Cyclone Nargis is now on the verge of that scenario. We are urging the international community to prevent that scenario by using all possible means to ensure that emergency humanitarian aid directly reaches to the people in need.

For long-time recovery for our impoverished country we need a transparent and responsible form of government responsive to the needs of the people. Democratic transition is the only solution for our country to overcome the challenges lying ahead.

Thank you, Mr. Chairman. I would be happy to answer your questions.

[The prepared statement of Mr. Win and material submitted for the record follow:]

PREPARED STATEMENT OF SEIN WIN, PH.D., PRIME MINISTER, NATIONAL COALITION GOVERNMENT OF THE UNION OF BURMA

Mr. Chairman,

Honorable Members of Congress,

Please allow me to express my sincere gratitude to the House Committee on Foreign Affairs, Subcommittee on Asia, the Pacific the Global Environment for organizing this hearing at this crucial juncture when my country, Burma, is facing a national calamity of an unprecedented nature in its modern history. The whole country and the people of Burma overseas are mourning for the millions of people affected by Cyclone Nargis when the storm hit the densely populated areas in Lower Bur ma on 2–3 May.

The recent cyclone has killed tens of thousands and ravaged homes, farms, and infrastructures leaving a lasting imprint of destruction which will take months, if not years, for the millions of survivors to recover from. Even though the death toll cannot be confirmed with certainty yet, the estimation so far is that over 100,000 people are dead or missing—40 percent of whom are believed to be children—and nearly 1.5–2.5 million people will suffer from the consequences of the cyclone.

Even under such circumstances, the Burmese military junta has done very little to help the people. Its main focus since the cyclone had been to hold a national referendum so that a constitution it had written to legitimize military rule would pass. At the same time it is also preventing the international community from entering the country with the fear that the rigid control that it has imposed on the country will be undone by the presence of international relief experts. In many areas, local authorities are also preventing local donors from visiting the cyclone victims and insisting that relief aid distribution must only be done by the authorities. What is even worse is that reports are trickling out about the international aid being embezzled by the authorities.

The Burmese generals' short-sighted policy has worsened the situation for the cyclone victims who are now exposed to torrential rains and strong winds brought on by the onset of monsoon in Burma. The weather is also making aid distribution even more difficult to the tens of thousands of survivors who are now living either in makeshift camps or monasteries or in their exposed storm-damaged homes which are surrounded by water filled with debris, and corpses and carcasses of draft animals. The death toll has been rising and with more survivors facing starvation, water shortage, and waterborne diseases, it can only increase. Safe water, food, and medical care need to be urgently provided now. An international NGO working inside Burma estimated that 30,000 children are starving and the under-fives living in the Irrawaddy Delta, who were already "acutely malnourished" when the cyclone hit Burma, might be dying from lack of food.

News agencies are also reporting that even in accessible areas like Kungyangon in the outskirts of Rangoon Division, men, women and children were standing in the mud and rain for miles alongside the road begging for scraps of food or clothing from an occasional passing aid vehicle. Tens of thousands of cyclone refugees are now crammed into monasteries and schools and most being fed and watered by local volunteers and private donors.

Our concern is that we are missing opportunities to help the victims and are risking more lives. Deaths this time around will not be from natural disasters but because of the xenophobic policies of the Burmese generals who are hindering international efforts. The situation is totally unacceptable and it must not be allowed to continue indefinitely.

Even though the Burmese generals have been dragging their foot, the international community has been responding very quickly. UN member nations have pledged US 79 million dollars to complement the 20 million dollars from the UN Natural Disaster Relief Fund to help Burma thanks to the flash appeal by the undersecretary for humanitarian affairs. UN humanitarian agencies and international humanitarian NGOs have mobilized their human and financial resources and are ready to go in. But, experts with relief expertise and experience from Western countries are being denied visas to enter the country. The junta, however, has allowed some international relief aid to be flown in and some Asian relief teams from nearby countries to enter the country. But the gestures are far too little given the magnitude of the problem the country is facing and they are attempts to ease international pressure and not genuine steps to help the people.

We have seen reports of naval ships from France and the United Kingdom loaded with food, medicines, safe water, speedboats, and helicopters arriving in the proximity of Burmese waters and preparing to help the cyclone victims. USS Essex and its battle group are also said to have been waiting to join in the relief efforts. U.S marine flights to Rangoon from their makeshift headquarters in Thailand's U-Tapao are continuing and a total of 500,000 pounds of aid were said to have been delivered. Negotiations to allow helicopters to fly directly to the disaster zone are being stalled and relief missions by French and British naval fleets are also waiting for permission from the Burmese military.

Information we are receiving from inside Burma indicates that effective distribution of aid is not in place yet two weeks after the cyclone. Access to the cyclone hit areas by the UN agencies and international humanitarian NGOs workers is still very challenging and many are still stranded in Bangkok waiting for visa. Meanwhile, the military regime has also been restricting the travel of non-Burmese staffs from UN agencies and INGOs already in Rangoon because the junta's Prime Minister Thein Sein has issued a "No foreigner, no camera" order to all military checkpoints.

Only Burmese staffs are allowed to travel to the cyclone hit areas. Even though the Burmese workers are willing and full of goodwill and compassion, they are sadly lacking in numbers and expertise. They are also not in the position to override local authorities when there is a need to make immediate decisions.

COMMUNITY INITIATIVE

The assistance regularly reaching cyclone victims so far have been from private donors, Buddhist monks, and celebrities. The National League for Democracy has also formed a natural disaster relief committee and starting to distribute rice, drinking water and clothing to cyclone victims through its network and NLD members, like others, are also encountering restrictions and interferences from time to time.

Many incidents of local officials using authority to take over relief aid packages have frequently been reported.

People have been sending word to us to help them regardless of the means involved.

RICE PRODUCTION AND FOOD SECURITY

The five regions struck by cyclone are predominantly agriculture zones where agriculture families producing 65 percent of the country's rice are living. The cyclone has now flooded rice fields with sea water and destroyed rice mills and storages as well as the salt and fishing industries (boats, fishing nets and storages). Farmers have also lost rice hoarded for family consumption, seed grains needed for the replanting season, as well as all their draft animals. The Food and Agriculture Organization has said that more than 20 percent of rice paddies in the cyclone-hit area were destroyed. Embankments and irrigation systems which form an intricate system critical to the success of rice crops have all been demolished and an enormous amount of work involving people who are grieving, homeless, and weak from hunger is needed to restore the destroyed infrastructures.

International agencies have time and again been warning that time is running out fast and if rice seed grains are not received within the next 40 to 50 days, planting will not happen in time for the harvesting this year and Burma will be subjected to the risk of famine if the agriculture system is not restored.

ABUSING AID

There have been reports of junta-appointed relief personnel abusing aid given by the international community as well as local donors.

In Bogale, a volunteer aid relief worker said local officials refused a private donor bringing in a truckload of relief goods. They demanded "one third" of whatever was on the truck as their share before the truck was allowed in.

In Insein, there was an incident of local authorities posing for the media with large quantities of rice and potatoes being distributed. But when the actual distribution is made, a cyclone victim is only given a potato and two cans of rice.

At the Kyongyi shelter for cyclone victims in Twante, a relief aid of three bags of rice, 1,000 eggs, two crates of tomatoes, two boxes of cigarettes, and two boxes of Thai instant noodles, Yum Yum, were donated to the cyclone victims in the name of Minister Soe Tha. After the minister left the area, two firefighters arrived at the shelter asking for half of what the minister had donated, saying that the share would go to the local authorities. The cyclone refugees refused to hand over the relief goods complaining that they did not receive any supplies donated by the minister when he came the first time round.

Incidents similar to the examples above are taking place everywhere. Under the situation, the Burmese generals have pledged to investigate any misuse of relief aid meant for cyclone victims and vowed to take tough action.

The only problem with that promise is finding people who will lead the investigations. Official corruption, from the lower echelons up, is widespread and appointing investigating officials, particularly at the grassroots, will not work because it is these very officials who are abusing relief aid.

RECOMMENDATIONS

1. OPENING UP AIR, SEA, AND LAND ROUTES FOR MASSIVE AID FLOW

The current flow of aid is way below what is needed to effectively save lives and relieve and rehabilitate cyclone victims. The amount of aid being delivered now [but not effectively distributed yet] is just one third of what is needed for the 1.6 to 2.5 millions of people affected by the cyclone. The military junta must open up every air, sea and land routes available to bring in a massive flow of aid to prevent deaths resulting from starvation and diseases.

2. EFFECTIVE DISTRIBUTION

Relief operations urgently need helicopters, speedboats, and shallow-water vessels. The infrastructure in place is insufficient to distribute international aid effectively to the needy. No modern facilities to unload relief goods from cargo planes exist. Roads have been damaged and ships only have limited access in the Irrawaddy Delta.

3. COORDINATING/MONITORING MECHANISM

A coordination mechanism is needed by the international community to ensure transparency and accountability. The UN Secretary-General has proposed a coordinating mechanism involving the UN and ASEAN. That mechanism also needs to be set up inside Burma and must involve major donor nations like the United States and EU countries. The military regime's relief committees established with local authorities and Union Solidarity and Development Association members are the current source of the relief problems: Relief aid from international organizations are relabeled as being given by the generals and/or ministers, relief goods are being replaced with defective or poor quality products, and relief goods are being stolen and sold in the local markets.

4. EXPERTS

International humanitarian experts, medical, and engineering teams are urgently needed to be on the ground now. Visa waiver for these relief officials, delivery and clearance of relief goods must be expedited, and free access must be given to disaster zones.

5. EMPOWERMENT OF CIVIL SOCIETY INITIATIVES

UN agencies and INGOs should work with local voluntary and civil society organizations as partners. Financial resources and technical skills should be extended and these informal civil society organizations should be empowered.

6. EXHAUSTING DIPLOMATIC OPTIONS

France took the initiative of seeking support from the UN Security Council to invoke the doctrine of responsibility to protect and we are grateful for that. Russia and China have consistently been opposing all efforts to get the international community to assist in Burma. The junta's supremo Than Shwe has not been receiving phone calls from UN Secretary-General Ban Ki-moon who we understand will try to go to Burma this week. Thai Prime Minister Samak Sundaravej conveyed a letter from the UN Secretary-General but failed to convince the junta to receive humanitarian experts.

7. HUMANITARIAN INTERVENTION

With diplomatic options quickly exhausting but urgent need for international aid growing in Burma, we believe that it is time for the international community to intervene to protect the distressed people of Burma. The leading political party—the National League for Democracy—and elected representatives in Burma, who are witnessing the calamitous situation, are also demanding an immediate humanitarian intervention.

What is obvious is that since the Burmese generals are willing to risk the lives of the people by preventing international humanitarian aid to be delivered to them, it has become the responsibility of nations and international agencies to rescue the needy in Burma regardless of the decision of the generals.

This can be done by nations with the right capabilities joining forces to enter Burma and save the lives of the millions who are at risk. Like minded countries can form an "International Coalition of Mercy" which will wholeheartedly be welcomed by an overwhelming majority of people of Burma, including the rank and file in the military. Since the Coalition is there to save lives and with the popular support of the people of Burma, we do not foresee China extending military support to the junta to counter the move. Besides, China is currently occupied with the recent earthquake disaster and the Olympic Games will not risk damaging its image or causing tension by siding with the Burmese generals whom the Chinese leaders know are way beyond redemption.

We, therefore, call for urgent humanitarian intervention by an international coalition of nations willing to help Burma. Diplomatic initiatives can be pursued following the entry into Burma by the coalition. Intervention not diplomacy needs to come first under the circumstances.

ANALYSIS OF THE SPDC'S REFERENDUM

The Burmese generals, ignoring widespread human suffering caused by cyclone Nargis and rejecting the appeal of the UN Secretary-General went ahead with the constitutional referendum in the midst of the humanitarian crisis. The referendum was held in the whole country on 10 May with the exception of townships affected by the cyclone in Rangoon and Irrawaddy Divisions.

The international community already knows about the flaws in the referendum law and the military's utilization of the state machinery, resources, and media as well as the use of intimidation, harassment, arrest, and physical attacks to ensure that the proposed constitution is approved by the referendum.

Even before the referendum was held, public service personnel and members of the armed forces, the Union Solidarity and Development Association, fire brigade and the national Red Cross were told that since they might have to travel on duty on the day of referendum, they needed to vote in advance. All these voters were made either to vote in the presence of or to hand over their completed ballot cards to the authorities, leading them without option but to support the constitution. The advanced voting took place in army battalions, government offices, and state-owned factories through out the country in April and May.

We have been monitoring the referendum and the following are some of our observations:

- The general atmosphere on 10 May was quite different from that of the general elections in 1990. People then were enthusiastic and long queues could be seen at the polling stations as people were eager to vote. A festive mood prevailed then because people believed that their votes would bring about democratic changes in Burma. Not too many voters were at the polling stations this time round and at times, there were more guards than voters. People had looks of anger, fear and desperation without displaying any enthusiasm. Team members report about many polling stations closing early around 11:00 am even though the official closing time is only at 04:00 pm.
- Even though the referendum law and rules and regulations governing polling station officials were promulgated by the military government, there were blatant violations of these legal rulings.
- Polling station officials in many areas were local authorities. Furthermore, members of the military-backed Union Solidarity and Development Association and/or National Red Cross Society who were not official polling station staffs were deployed around poll stations. From time to time, they were seen to be accompanying voters into the polling booths and guiding them to vote "Yes."
- In many regions, local authorities went house to house the night before the referendum to collect advanced votes from the local people. Voters were forced to tick 'yes' on the ballot in their presence. Polling stations were closed early and those came to vote were asked to either vote at the offices of the local authorities or return home because their votes had already been collected in advance.
- Local authorities retained national identity cards and bank books from farmers (the bank books are needed for borrowing agricultural loans from government banks). The IDs and banks were returned to the farmers only after they had ticked "yes" on the ballot cards in the presence of the local authorities at the local Peace and Development Council offices.
- Polling station staff members were seen handing out ballot papers that had already been checked in favor the constitution. Voters were compelled to sign to register their presence and their pre-voted ballot cards placed into ballot boxes in front of authorities.
- In some areas only one member of the household was asked to come to the polling station and forced to vote "yes" for all eligible voters in the family.
- Local authorities went around village/town announcing through loudspeakers that those who voted against the constitution would face legal action and imprisonment of up to 3 years and a fine of 10,000 Kyats. Misinformation campaign about such legal action was launched in some regions to create fear among the uneducated and not well informed people in rural areas.
- In many areas, NLD members insisted on joining the vote counting process as observers as required by law but they were denied. Heated arguments ensued and NLD members filed official complaints to the higher level referendum sub-commissions about the violation of referendum law and regulations.
- Thakin Soe Myint, chairman of the Rangoon Division NLD Organizing Committee, and Joint Secretary Dr. Myo Aung were prevented from traveling to Hmawbi, Taikkyi and Tantabin Townships to observe the referendum. They were stopped by police at a tollgate near Hmawbi and forced to return to Rangoon without giving any reason. Three NLD youth members who informally observed the referendum were also arrested.
- The referendum organized on 10 May was neither free nor fair and it violated all internationally accepted norms. The Burmese military and its agents manipulated the process using all available means. The Burmese generals are

expected to announce an "overwhelming" result in their favor which was achieved through extensive fraudulent practices.

US POLICY ON BURMA

The military junta has announced that its constitution has been approved by the referendum and the process to legitimize military is already in the works. We foresee even greater political instability in the coming months as people are bound to be frustrated with the worsening socioeconomic crisis caused by the effect of cyclone Nargis and the regime's lack of regard for human suffering. Given the fact that the referendum did not reflect the will of the people since it was methodically manipulated by the military, I strongly urge the United States to reject the outcome of referendum and the military junta's constitution.

The United States has always been one of the strongest supporters of Daw Aung San Suu Kyi and the Burmese democracy movement and is consistent in its unwavering support for the efforts to bring about peaceful democratic transition and national reconciliation in Burma. The people of Burma have always been grateful to the United States for its support.

I would like to appeal to the United States to continue taking the lead at the international level, effectively coordinating with United Nations, the European Union and ASEAN to push for speedy democratic reforms in Burma and a UN Security Council resolution whenever possible. The United States is also urged to help the people of Burma in the following policy areas:

Increase Pressure on the Regime and Taking a Lead in Mobilizing Policies toward Burma: US Government should take new policy initiatives in addition to existing policy measures such as the strengthening of banking and financial sanctions targeting family members of the regime and their business cronies until a negotiated political settlement for a complete transition to democratic regime becomes irreversible.

Greater US initiative, engaging with China, India, and ASEAN: Since neighboring countries like China, India, and ASEAN are a concern because of their lenient and friendly policies toward the Burmese regime, U.S. Administration and congressional leaders are requested to continue bringing Burma issue to the fore in bilateral talks with China, India, and ASEAN as well as at ASEAN post ministerial meetings and ASEAN Regional Forum.

Increased humanitarian assistance: In response to the humanitarian crisis in Burma, US Government is requested to increase humanitarian assistance to the people of Burma, including the delivery of aid across borders to refugees and IDPs, through credible international NGOs and informal civil society groups. Strengthening the capacity of informal civil society groups can be done by Burmese exile groups which can organize short term courses, workshops, exposure trips and internship program with experienced NGOs in neighboring countries.

Taking a Holistic Approach to Cooperation in Drug Eradication: US Government is urged not to de-link drug cooperation from the larger policy goal of promoting democracy in Burma since drug problems in Burma will remain unresolved until the underlying political and economic problems are settled in a democratic way.

Empowerment of the Burmese Democratic Forces: Leading Burmese democratic forces like the NLD and the government in exile, National Coalition Government of the Union of Burma, place high importance on preparations toward post-military transition in Burma and the United States is asked to increase support to the democratic forces in strengthening their efforts for the restoration of democratic governance, national reconciliation, and civil society in Burma.

INTERNATIONAL
Herald Tribune

The junta's vote for repression

By Sein Win

Monday, May 5, 2008

It is not often that there is fair warning of a comprehensive abuse of political rights. But in Burma, we know the day that this abuse will take place: This coming Saturday. This is the day of the referendum called by Burma's military regime on the draft constitution it has prepared.

The preparation of the draft constitution has been anything but democratic and inclusive. The process has excluded all democratic groups and parties, including the National League of Democracy, which won Burma's last democratic elections in 1990. It has also excluded the legitimate representatives of Burma's many ethnic groups.

The constitution proposed by the regime - the State Peace and Development Council - is one that would leave the military in power, regardless of the wishes of the Burmese population.

The elections that the regime have proposed to follow the referendum in 2010 would not permit the participation of many democratic parties and individuals, including the leader of the National League for Democracy, Aung San Suu Kyi, who remains under house arrest, now for over 12 years.

In order to view the draft constitution, Burmese citizens - who, thanks to the regime's disastrous economic policies, are among the poorest in the world - are required to purchase the 194-page document. Already, the ruling junta is organizing acts of intimidation and violence to bully the population to vote yes to the constitution. Despite the obvious risks to their lives and liberty, opposition groups have decided to participate in the referendum and vote no.

The regime wants to present this false referendum and anti-democratic constitution as part of a transition to democracy. The reality of the referendum is a cynically calculated act to pretend to meet the demands of the international community, including the UN Security Council, which has specifically demanded an inclusive constitutional process.

Since the violent repression of demonstrations led by Burma's revered monks in September, the UN has conducted a "good offices" mission to try to find a way forward in Burma. So far, the ruling junta has shown no sign of permitting any kind of inclusive constitutional and democratic process. Their tactics seem to be to stave off international pressure by slowing down any dialogue with the UN, and promising empty changes, such as its new constitution.

These evasive tactics must not be tolerated. The UN Security Council, along with Asean, the European Union and Burma's immediate neighbors, must call this referendum what it really is and demand an inclusive constitutional process, leading to genuinely democratic and open elections.

These countries - the international community - must insist on objective international monitoring of the referendum - perhaps by Burma's democratic regional neighbors like Thailand, Malaysia, Singapore, or Indonesia - and, if this is refused, draw the obvious conclusion.

It is time for the international community to increase the economic pressure on the military senior leadership via the mechanism of targeted sanctions, including investment freezes. Those countries in Asean, as well as China and India, which are the military's major trading and investment partners, must cease propping up the regime with currency, revenue and arms.

It is not in anyone's long-term interests for a regime that lacks any legitimacy to continue in power in Burma. No economic investment can prosper in such circumstances and the risk of widespread turmoil and instability can only increase. Above all, the abuse of the human rights of Burma's people must be stopped.

The country's democratic opposition proposes a way out of this crisis and suffering for the Burmese people. We are not looking for revolution or violent overthrow of the regime, but a peaceful transition to a situation where Burma is stable and democratically governed.

What this requires is a constitutional process that includes all parties, including the military, democratic parties and representatives of the many ethnic groups that make up contemporary Burma. All these groups must have a say in Burma's constitutional settlement, as they must too play a part in any future government. The proper atmosphere for such a process also requires the release of political prisoners and the end to hostilities in Burma's ethnic regions.

The military regime pretends that the referendum is a step toward democracy. It is in reality a massive and comprehensive denial of the democratic and political rights of the Burmese people. It is essential that the international community recognizes this fake referendum for what it is.

Sein Win is chairman of the National Coalition Government of the Union of Burma, which is outlawed by the military junta.

After the Storm

By Thaung Htun

As Burma reels under the deadly impact of Cyclone Nargis, the referendum on the military government's sham constitution, set for this Saturday, needs to be postponed. Doing so will not only allow Burma to begin to recover, but will provide a moment to consider the many myths surrounding the military's self-centered agenda.

Cyclone Nargis creates a much-needed pause for reflection in Burma.

The first myth is that Burma needs a strong constitution, infused with military muscle, to ensure the country does not descend into anarchy. Because of Burma's strategic location, its eight major ethnic groups and its access to plentiful natural resources, neighboring countries, the logic goes, fear a "Balkanization" of Burma.

That's simply untrue. Virtually all ethnic groups have united against the military regime. One example of this cooperation is the Committee Representing the People's Parliament, formed in 1998, which includes representatives of the National League for Democracy and two of Burma's major ethnic alliances: the United Nationalities' League for Democracy and the United Nationalities Alliance, the umbrella group of all ethnic-based political parties.

The National Coalition Government of the Union of Burma (for whom this author works) and ethnic parties have also achieved important accords, which formalize agreements to solve any differences among ethnic groups by means of peaceful negotiations and on the basis of equality. Similar alliances forged in exile have also worked on a future constitution through a broad-based Federal Constitution Drafting process and have reached agreement on fundamental principles.

The last time Burma held democratic elections, in 1990, they were anything but divisive. The National League for Democracy and the United Nationalities League for Democracy together won 95% of parliamentary seats. This suggests overwhelming support for a united approach in Burma, and confirmation that so-called "ethnic tensions" are largely fallacious.

Another myth is based on the mistaken belief that there is no democratic infrastructure in Burma. This theory holds that the military is the only option for Burma because there is nothing to replace it. This argument completely overlooks the capacity of the Committee Representing Peoples' Parliament, which obtained a mandate from the people through the 1990 general elections and has developed transitional policies on key agendas to reconstruct the country. However, the parliament was never allowed to convene inside the country and form a government.

Perhaps the most credible confirmation of a democratic alternative in Burma is the staying power of the ideals of Aung San Suu Kyi, who heads the National League for Democracy. Despite being incarcerated by various means for 12 of the last 14 years, Ms. Suu Kyi has led our people on the basis of free thinking, free association, nonviolence and the principles of democracy.

A third myth is that the new draft constitution and the parallel promise of full elections in 2010 are signs that the military leaders are opening up to democracy. This is belied by the fact the military has had every opportunity to work toward democratic values since it took power in the coup of 1962, and has not done so. The military not only disqualified the democratically elected government after elections in 1990, it also has never offered any explanations as to what role the government should play in a democratic transition.

The very process of writing this constitution, some two decades in the making, ensures the continued abrogation of democracy. The document seeks to set in stone a 25% representation of the military in any parliament. It has been drafted without any public consultation, and was released for sale only weeks before the referendum. Voters in Burma are being intimidated and pressured to vote "yes," while those outside the country and are having restrictions placed on their ability to cast votes freely.

This is no democratic process. This is authoritarianism donning the suit of democracy, for a brief appearance in the spotlight, before returning to its olive drab garb. No one should be taken by this ruse, nor allow the awful consequences of Cyclone Nargis to be an excuse to ignore the equally dire reality of Burma's regime.

With Nargis's death toll rising and people desperately searching for shelter, food and water, the country should not be asked to consider such a weighted proposition as Saturday's referendum. This is no time for politics. It is a time for reflection and for the international community to not only offer aid, but to reconsider the myths surrounding Burma.

Mr. Thaung Htun is the representative for United Nations Affairs for the National Coalition Government of the Union of Burma, Burma's government-in-exile.

THE WALL STREET JOURNAL.

Published since 1889 by
DOW JONES & COMPANY.
DOW JONES
A NEWS CORPORATION COMPANY

Mr. FALEOMAVAEGA. Thank you, Mr. Prime Minister.
Congressman Andrews, for your statement.

STATEMENT OF THE HONORABLE THOMAS H. ANDREWS, PRESIDENT, NEW ECONOMY COMMUNICATIONS (FORMER MEMBER, U.S. HOUSE OF REPRESENTATIVES)

Mr. ANDREWS. Thank you very much, Mr. Chairman, and thank you so much for holding this public hearing this morning. It is good to be back but unfortunately under these circumstances.

Mr. Chairman, even more horrifying than the ravages of the cyclone that hit Burma more than 2 weeks ago is the cruel and devastating catastrophe that continues to claim the lives of tens of thousands of victims at the hands of the brutal military dictatorship of Burma. It is estimated that 2.5 million are in dire need of shelter, clean water, food and medical treatment right now. The refusal of the military Government of Burma to allow life-saving relief into the country while confiscating and reportedly selling at least a portion of the trickle of aid that has been allowed in is nothing short of criminal. For growing numbers of citizens of Burma, many of whom are children, the reprehensible actions of the military junta of Burma constitute a death sentence. These acts are clearly a crime against humanity.

While horrifying and reprehensible, the actions of the military junta of Burma following Cyclone Nargis come as no surprise to anyone familiar with this regime. Since the regime violently crushed a mass uprising in Burma in 1988, it has brutalized its people. Since 1962, the military-run Government of Burma has decimated its once promising economy, looted its vast natural resources and destroyed vital service sectors, such as health care. Today, Burma's health sector ranks 190th out of 191 nations. Health care receives 3 percent of the regime's annual budget compared with the 40 percent dedicated to the military.

International organizations, including the United Nations, have documented year after year the atrocities of this regime from the forced labor of its workers to the destruction of entire villages to the systematic rape of thousands of women and girls. The brutalization of the people of Burma has been particularly egregious in eastern Burma where a scorched earth campaign has destroyed or forced the abandonment of more than 3,000 villages. To put this into context, Mr. Chairman, this is twice as many villages as has been destroyed in Darfur.

Then as now the international community has sought to take action to defend the defenseless victims in Burma largely through the United Nations. Then as now the regime has resisted. When governments felt compelled to respond to the despicable and inexcusable acts of the military government, the regime would announce a new policy. But time again and again, it would become very clear that these new policies were designed only to give the appearance of reform so as to reduce international pressure.

The regime's so-called disciplined democracy with the sham elections we have witnessed are a good example. These so-called elections were part of a process that was initiated after the international outcry over the brutal suppression of the democracy movement of Burma, including incarceration of Nobel Peace Laureate

Aung San Suu Kyi, duly elected members of Parliament and democracy activists. While the public announcement that the military junta has agreed to an ASEAN-led task force for redistribution of foreign aid might sound like a breakthrough, there is every reason to believe and fear that it will be the latest manifestation of the regime's systematic use of manipulation and deception to ease mounting pressure from the international community.

The early signs are not good, as they follow a clear pattern. It appears now that while it has been announced that the generals ruling Burma have agreed to accept an ASEAN Emergency Rapid Assistance Team, they are in no hurry to allow critically needed humanitarian experts to enter the country. In the very same Associated Press story announcing this breakthrough, the foreign minister of Singapore, which convened the emergency ASEAN meeting, said that the junta had not agreed to allow any aid experts to arrive immediately. Immediately, of course, is exactly what is desperately needed to save tens of thousands of lives.

Even more distressing is the fact that many of the aid workers who have been allowed into Burma have not been allowed to do their job. An official of an ASEAN country told me very recently that emergency response team members that have been allowed into Burma from his country have spent precious days sitting in Rangoon waiting for permission to get to work. With any agreement with the ruling generals of Burma, the devils are in the details. In fact, I cannot think of a situation where this phrase is more appropriate.

It is imperative that aid workers with critical experience in dealing with these types of natural disasters, including those who coordinated and delivered aid to the tsunami victims in 2004, be allowed into Burma immediately and give it unfettered access to the Irrawaddy Delta region.

It is encouraging that the Secretary-General of the United Nations is willing to travel to Burma and is being allowed into the country. It has taken weeks for his phone calls to be answered. But the Secretary-General needs the full support the international community, including the United States, to assure that what emerges from these negotiations is a desperately needed breakthrough that will save lives and not a ploy merely to give appearances of change.

Then as now the regime has sought and found relief from international pressure from countries who, in the name of national sovereignty, become enablers and accomplices to the regime's brutality. No nation has played a more powerful role in this regard, Mr. Chairman, than China. And I would like to emphasize this point. China continues to serve as the military regime's number one supplier of weapons and military equipment, enabling the regime to amass one of the largest armies on the planet now exceeding 400,000 troops, 70,000 of which are children. These were the weapons that were used last to violently crush the peaceful demonstrations of Burmese monks last fall. China has consistently protected the regime from international pressure casting or threatening to cast vetoes in the United Nations Security Council that would bring pressure to bear on the military government.

In exchange the generals supply China with deeply discounted supplies of natural gas and other natural resources often, as documented by the international labor organization, extracted and transported with the use of forced labor. China is Burma's number one supplier of imports with trade revenues in the billions of dollars. It invests heavily in a large number of companies owned and controlled by the military junta. Then as now China led opposition to any kind of international action. It has opposed even the consideration of United Nations Security Council action to confront the overwhelming evidence that we are watching what amounts to the mass murder of untold millions of innocent people living in the Irrawaddy Delta.

There are some important lessons that the international community needs to draw from the bitter experience of this regime Mr. Chairman. First, the ruling regime in Burma is a ruthless and treacherous regime that will do anything to maintain its iron grip on power. Second, the regime is susceptible to international pressure, but facing such pressure, it will employ deception and manipulation to give the appearance of change, decide not to bring reform to Burma but relief to the ruling generals. And finally, third, Mr. Chairman, the regime's greatest protector and enabler is China. From supplying vast amounts of military weapons, equipment and training to its investments in the junta-owned enterprises, to its vetoes on the United Nations Security Council, China can be counted on as an accomplice to the generals' treachery. Yes, the ruling generals have been able to rely on their ASEAN neighbors to go along with its deceptions and hide behind the banner of national sovereignty, but there is not greater enabler of the ruling military junta of Burma and therefore no greater source of influence than China.

Mr. Chairman, I would submit the rest of my testimony for the record. I know that time is precious, but I thank you once again for your willingness to convene this hearing. I would be more than happy to answer any questions.

[The prepared statement of Mr. Andrews follows:]

PREPARED STATEMENT OF THE HONORABLE THOMAS H. ANDREWS, PRESIDENT, NEW ECONOMY COMMUNICATIONS (FORMER MEMBER, U.S. HOUSE OF REPRESENTATIVES)

Mr. Chairman and Members of the Committee,

Thank you for holding this public hearing on the horrifying human tragedy that continues in the cyclone ravaged nation of Burma and thank you for the opportunity to join you today.

My name is Thomas Andrews, I am former member of the House of Representatives (D–ME) and president of New Economy Communications, a not-for-profit organization that provides consultation, strategic planning and communication services to human rights, labor rights and democracy promotion organizations in the United States and abroad.

I am also a Senior Advisor to the National Democratic Institute for International Affairs and provide assistance to the National Coalition Government of the Union of Burma, led by Prime Minister-in-Exile, Dr. Sein Win, who is here today, as well as an international network of organizations who advocate for a free and democratic Burma.

Even more horrifying than the ravages of the cyclone that hit Burma more than two weeks ago, is the cruel and devastating catastrophe that continues to claim the lives of many tens of thousands of victims at the hands of the brutal military dictatorship of Burma. It is estimated that 2.5 million are in dire need of shelter, clean water, food and medical treatment. The refusal of the military government of Burma to allow life-saving relief into the country, while confiscating and reportedly selling at least a portion of the trickle of aid that has been allowed in, is nothing short of criminal. For growing numbers of citizens of Burma, many of whom are children, the reprehensible actions of the military junta of Burma constitute a death sentence. These acts are clearly a crime against humanity.

While horrifying and reprehensible, the actions of the military junta of Burma following Cyclone Nargis come as no surprise to anyone familiar with the regime. Since the regime violently crushed a mass uprising in Burma in 1988, it has brutalized its people. Since 1962, the military run government of Burma has decimated its once promising economy, looted its vast natural resources and destroyed vital service sectors such as health care. Today Burma's health sector ranks 190 out of 191 nations. Health care receives three percent of the regime's annual budget, compared with the 40% dedicated to the military.

International organizations, including the United Nations, have documented, year after year, the atrocities of this regime from the forced labor of its workers, to the destruction of entire villages to the systematic rape of thousands of women and girls. The brutalization of the people of Burma has been particularly egregious in Eastern Burma where a scorched earth campaign has destroyed or forced the abandonment of more than 3,000 villages. To put this in context, this is twice as many villages as have been destroyed in Darfur.

Then, as now, the international community has sought to take action to defend the defenseless victims in Burma largely through the United Nations. Then, as now, the regime has resisted. When governments felt compelled to respond to the despicable and inexcusable acts of the military government, the regime would announce a new policy. But, time and again, it became very clear that these new policies were designed only to give the appearance of reform so as to reduce international pressure. The regimes so-called "disciplined democracy" with the sham elections we have witnessed are a good example. These so-called elections were part of a process that was initiated after the international outcry over the brutal suppression of the democracy movement in Burma, including the incarceration of Nobel Peace Laureate Aung San Suu Kyi, duly elected members of parliament and democracy activists.

While the public announcement that the military junta has agreed to an ASEAN-led task force for redistributing foreign aid might sound like a breakthrough, there is every reason to fear that it will be the latest manifestation of the regime's systematic use of manipulation and deception to ease mounting pressure from the international community.

The early signs are not good as they follow a clear pattern: it appears now that while it has been announced that the generals ruling Burma have agreed to accept an ASEAN Emergency Rapid Assessment Team, they are in no hurry to allow critically needed humanitarian experts to enter the country. In the very same AP story announcing the breakthrough, the Foreign Minister of Singapore, which convened the emergency ASEAN meeting, said that the junta had not agreed to allow any aid experts to arrive "immediately." "Immediately," of course, is exactly what is desperately needed to save tens of thousands of lives. Even more distressing is the fact that many of the aid workers who have been allowed into Burma, have not been allowed to do their job. An official of an ASEAN country told me that emergency response team members that have been allowed into Burma from his country have spent precious days sitting in Rangoon waiting for permission to get to work.

With any agreement with the ruling generals of Burma, the devil is in the details. In fact, I cannot think of a situation in which that phrase is more appropriate. It is imperative that aid workers with critical experience in dealing with these types of natural disasters, including those who coordinated and delivered aid to tsunami victims in 2004, be allowed into Burma immediately and given unfettered access to the Irrawaddy delta region. It is imperative that the aid and equipment sitting offshore on American, French and British ships be deployed immediately. And, it is crucial that all aid entering the country be monitored closely so that it goes to the people of Burma who are desperately in need. Unfortunately, none of these desperately needed steps are part of this agreement and the international aid conference for Burma that has been announced will not even begin until next week.

It is encouraging that the Secretary General of the United Nations is willing to travel to Burma and is being allowed into the country. It has taken weeks for his phone calls to be answered. But the Secretary General needs the full support of the international community, including the United States, to assure that what emerges from these negotiations is a desperately needed breakthrough that will save lives and not a ploy that merely gives the appearance of change while taking the pressure off.

Then, as now, the regime has sought and has found relief from international pressure from countries who, in the name of "national sovereignty" become enablers and accomplices to the regime's brutality. No nation has played a more powerful role in this regard than China.

China continues to serve as the military regime's number one supplier of weapons and military equipment, enabling the regime to amass one of the largest armies in the world, now exceeding 400,000 troops—70,000 of whom are children. These were the weapons that were used last fall to violently crush the peaceful demonstrations of Burmese monks. China has consistently protected the regime from international pressure, casting or threatening to cast vetoes in the United Nations Security Council that would bring pressure to bear on the military government. In exchange, the generals supply China with deeply discounted supplies of natural gas and other natural resources (often, as documented by the International Labor Organization, extracted and transported with the use of forced labor). China is Burma's number one supplier of imports with trade revenue in the billions of dollars. It invests heavily in a large number of companies owned and controlled by the junta.

Then, as now, China led opposition to any kind of international action. It has opposed even the consideration of United Nations Security Council action to confront the overwhelming evidence that we are watching what amounts to the mass murder of untold numbers of innocent people living in the Irrawaddy delta.

There are some important lessons that the international community needs to be drawn from bitter experience with the military regime of Burma:

First, the ruling regime in Burma is a ruthless and treacherous regime that will do anything to maintain its iron grip on power. Even before it was hit by a massive cyclone, untold numbers of people in Burma have lost their lives at the hands of the military regime's brutality.

Second, the regime is susceptible to international pressure, but, facing such pressure; it will employ deception and manipulation to give the appearance of change designed not to bring reform to Burma, but relief to the ruling generals.

Third, the regime's greatest protector and enabler is China. From supplying vast amounts of military weapons, equipment and training, to its investments in junta owned enterprises, to its vetoes on the UNSC, China can be counted on as an accomplice to the general's treachery. Yes, the ruling generals have been able to rely on their ASEAN neighbors to go along with its deceptions and hide behind the banner of "national sovereignty," but there is no greater enabler of the ruling military junta of Burma—and therefore no greater source of influence—than China.

I believe that the United States government needs to make the crises in Burma a much higher priority. A high ranking U.S. official must be appointed and empowered to take the lead in this crises, working tirelessly within the diplomatic community to muster the necessary support for action. Secretary of State Powell played a key leading role when the tsunami hit Southeast Asia in 2004, travelling and communicating directly and often with his counterparts from nations around the world and leading a robust and multi-faceted approach. A President's Special Coordinator for International Disaster Assistance was established to streamline and coordinate the efforts of US government aid and assistance. We did that to confront one major disaster—the tsunami. With Burma we are confronting two disasters—the devastation wrought by a powerful cyclone and the devastation wrought by a murderous regime willing to sacrifice the lives of many tens of thousands of its citizens in order to keep those who survive under its brutal thumb. In order to be able to deliver a coordinated aid effort to save thousands, the United States must pursue an aggressive and relentless campaign to build international pressure on the regime and those regimes that protect it.

The single most important target of that pressure must be China. While China has joined the international chorus calling on Burma to allow aid into the country, it is hardly enough. It is hard for me to believe that if China were to put serious and sustained pressure on the military regime of Burma, change would follow and tens of thousands of lives would be saved. After all, in so many ways, the regime owes its very existence to its patrons in the Chinese government.

Relentless international pressure is needed to save the hundreds of thousands whose lives are hanging in the balance at this very moment.

We know what to do. An air bridge of relief that sent aircraft filled with supplies and experts into Aceh every hour during the 2004 tsunami was set up within 72 hours of that disaster.

But first there must be the international will to act.

Thank you.

Mr. FALEOMAVAEGA. Thank you, Mr. Andrews. Congresswoman Watson for her questions.

Ms. WATSON. I think your last paragraph is the most important one for this committee. It a recommendation that we set a priority on this, and I couldn't agree more. Because many of us are unfamiliar with the way this administration works in terms of foreign policy, who would you suggest that special envoy be, at what level? Should we work it through the U.N.? And I just heard that the Secretary-General of the U.N. is on his way. Is he someone we should work through? Should we send our own envoy? Would you give us your best thinking on that, Mr. Andrews?

Mr. ANDREWS. Thank you. I think a good example is looking at the tsunami crisis in 2004 in which the administration at the very highest level, Secretary Powell, took a very hands on, very public role and also a very private role by traveling extensively, communicating daily, nightly to coordinate the messages to make certain that there was full coordination and support from the United States. That was the highest level of our State Department obviously, and I think that that was a reasonable response.

We also established a Special Coordinator for International Disaster Assistance to make sure that all the various branches of our own Government were acting in concert. I think that was extremely important as well.

This needs to be conducted at the highest level. We all have different roles to play as nations, as you well know, but I think if the United Nations plays a most critical role that the United States, using all of its diplomatic tools and resources, can help the United Nations succeed.

Ms. WATSON. Mr. Chairman, as a committee I think we ought to make this strong recommendation. I don't know what our Secretary of State is doing in this part of the world. I get the feedback from the Middle East and maybe Africa, but I think we ought to make a strong effort if we ourselves cannot get not visas to go, that for sure our Secretary of State put the hands on this situation.

As we watch TV it is like the thinning out of their population on purpose, which is a criminal act as far as I am concerned. At least China is doing everything they can for the earthquake victims. I really have greater respect for how they have moved in to help the lowliest and the innocent. But I think this situation in Burma cannot continue on without top level. So I would like us to contact the administration, Mr. Chairman, at your will, but let them know what priority this committee sees on the way they are handling the situation and they need to be more involved at the top.

Thank you, Mr. Chairman. And I must leave.

Mr. FALEOMAVAEGA. I thank the gentlelady. The gentleman from California.

Mr. ROHRABACHER. Thank you very much. And Dr. Win, it is good to see you again and over the years we have been active in so many efforts and now we are here discussing this tragedy.

Dr. Win, I would take it that you disagree with the chairman's remarks that perhaps the United States should not expect to be cooperated with, seeing that we have been so antagonistic toward the junta in Burma.

Mr. FALEOMAVAEGA. Will the gentleman yield?

Mr. ROHRABACHER. Certainly.

Mr. FALEOMAVAEGA. The point that I wanted to make on that, as I stated earlier, just as a point of observation, not necessarily disagreeing with our contention or saying that Burma does not have a democratic government, but I am talking about the perception. The fact that we condemn Burma on the one hand and then on the other hand we want to give humanitarian assistance—I think in the minds of our people there, our friends and, if you want to call them friends, the military regime there, how can you expect us to receive your assistance when you are trying to kill us?

Mr. ROHRABACHER. Well, I would think that is a perception only. That is only a problem with the perception of these gangsters, and who cares what their perception is.

Dr. Win, do you think the people of Burma believe that we have been too antagonistic and that is sort of one of the reasons that we are having trouble now getting relief, or is it squarely on the shoulders of these somewhat evil gangsters who run that regime for the fact that they perceive it that way?

Mr. WIN. Yes, when we talk about the regime, their priority is not to help the people but to control the people. So for them the suffering of the people is not in their agenda. So what they are afraid is that people will see what the real situation on Burma, what they are afraid if they were to lose the control, and that is why they are now reacting in this way. So as all the time they are not only not letting the United States to come in, but they also have not reacted positively to the United Nations. And also even not to the ASEAN until now the pressure is on them, a huge pressure is on them.

So we don't think this is only with the United States, they don't also let European aid to come in.

Mr. ROHRABACHER. I think it is important for us to note and for the people of Burma to understand that their suffering is not caused by the natural disaster. All countries will experience natural disasters. And what we have now is suffering caused by an iron fisted dictatorship that rules their country. And I would disagree with any analysis that would suggest that the United States should refrain from being critical and proactive in trying to oppose such regimes in order to make sure that our relief efforts in case there is an emergency are facilitated.

In fact this is, as I pointed out, in some other countries like Nicaragua where the Somoza regime was a dictatorship as well and also mishandled the assistance and interfered with the assistance that was going to their people after a big earthquake that led to the destruction, not of the country of Nicaragua, but of the regime that controlled Nicaragua, the Somoza regime.

Dr. Win, do you believe that it is possible that this tragedy that the people of Burma have had to suffer could open the eyes of the people to the point that they must now eliminate this terrible burden that they have had to carry, this dictatorship? The young people in the military, perhaps their eyes may be opened that they should not be supporting this regime, a regime that gets in the way of humanitarian help at a time of crisis?

Mr. WIN. Yes. One thing about this is that this kind of cyclone destruction and the military failure to fulfill their responsibility have great effects on the whole population. I think even the military personnel family may have the same suffering because of this cyclone. So we definitely think that this will have a long lasting effect and consequence for this regime.

Mr. ROHRABACHER. Let me note for the record, and I know my time is up, it is just that what we are talking about is not politics. Here in the United States politics gets in the way of things, it does. Politics gets in the way. And we have differences of opinion and people are maneuvering through the democratic process. There is a big difference in politics getting in the way and evil getting in the way. What is getting in the way of assistance to those people who are suffering the results of the cyclone in Burma is evil, and that evil is personified in the generals who control that country with an iron fist. It is not political considerations that you are talking about, like in a democratic society.

Are we going to get a number of votes here? Are we going to be able to have this group support us here in terms of the next election? No, what we are talking about there is how strong our control is going to be in the future over the people of this country. That is an evil consideration, not a political consideration. And one only has to look back to after the Second World War when there is tremendous suffering in Europe and the Soviet Union at that time turned down the option of being part of the Marshall Plan in which the United States stepped forward and helped rebuild war torn Europe. But yet the Soviet Union and Communist regimes refused to let the United States even come in to help rebuild the torn countries.

This is very similar to what is going on in Burma and it is based on evil and not simply politics as we understand it in a democratic context.

So with that said, Dr. Win, good luck and let's hope within a short period of time the people of Burma's suffering will be alleviated both from the cyclone and their dictatorship.

Mr. FALEOMAVAEGA. Dr. Win, I wanted to share with you, you are probably very familiar with the history of this country itself. We were a colony of the British Empire. And what did we have to do eventually? We had to establish a revolution. We had to take it to battle to fight the mighty British Empire. What I wanted to ask you is if for 36 years you have been under this military regime, what are you hoping for? Are you advocating overthrowing the current military regime by force or are you hoping someday they will just dissipate and die off?

Mr. WIN. Well, as you know, this encourages us very much. And this, after cyclone, what we are talking about the consequence could be in that direction.

What I am wanting to say is that because, as you said, '88 and then Aung San Suu Kyi and Nobel prize, all these have some effect on our thinking about negotiation and also about dialogue, going to give them a lot of chance and flexibility for the sake of the country. But now with the Cyclone Nargis, maybe other option will be closed. So this is what I am trying to express.

Mr. FALEOMAVAEGA. I am not sure I am clear on what you are trying to say here, Dr. Win. I will say I don't think there is any disagreement between me and my good friend from California. The fact that there is an absence of democracy for 36 years, your people have suffered under the ruling military regime. I am asking this question for which my friend raised the issue, how do you want to get rid of evil if you describe the military regime as an evil empire?

What options do you have? Just putting pressure on the international community and pray and hope someday that the military regime is just going to say, okay, we give up? Do you think this is the way it is going to happen?

Mr. WIN. No, not exactly. What we are hoping is that the countries like the United States, which is the biggest democracy in the world, will lead, for example, the kind of emergency situation like this.

Mr. FALEOMAVAEGA. I am not talking about emergency. I am talking about what my friend is saying here. If we have an evil empire here, if you want change, how do you go about providing that change?

There is no question about humanitarian aid. Everybody knows it has been delayed. The military government has very serious questions about the motives of certain countries that want to bring assistance to Burma. That is the point that I am raising here.

Now my friend here says that Burma is an evil empire. I am asking you, What is your recommendation to get rid of the evil empire?

Mr. WIN. We have to use every means. That means also international pressure and that means also internal pressure.

Mr. FALEOMAVAEGA. But you have been doing this for the last 36 years.

Mr. WIN. Yes, it is. But even if you don't see a real difference, there is some difference in environmental. We know that because we have been there for many years in the silo, and when you compare this regime and let me say North Korea. You cannot control totally the population. The population is more flexible, I mean more opposition. You don't see a kind of demonstration in North Korea or opposition party or any kind of dissidence expressing, but in Burma you will see all those things.

So it is advancing slowly and our thinking is that we have to use every means to get rid of them. If they are willing to talk with us, that is good, because it will save the suffering and continue about the country. But we have to use other means and we hope to use all the means.

Mr. FALEOMAVAEGA. So what other means are you suggesting that we ought to do to get rid of these guys?

Mr. WIN. Well, no, kind of internal like, you know, or defiance and people powers, you know, and that kind and outside also we are trying to put pressure on the military, you know?

Mr. FALEOMAVAEGA. We talk about democracy, democracy is perceived in very different ways, just as democracy is trying to come out of Afghanistan, out of Iraq. Should it be perceived in the way that Americans want democracy operating or should it be done Burmese style? Which I don't know what that style may be. Do you think there is perhaps a little misunderstanding here or miscommunications or a lack of communication in how the military regime is conducting its business and its relationship to you after being elected and you were unfortunately taken out? I am trying to search here for an answer as to why after 36 years there still has not been any reconciliation between the current military regime and persons such as yourself who was duly elected as Prime Minister.

Congressman Andrews.

Mr. ANDREWS. Well, Mr. Chairman, there is an old saying he who pays the piper calls the tune. If you begin to look at how this Government of Burma is able to sustain itself, you will look at some of its major suppliers of weapons for the largest army in the world, for its economy, and that clearly has to focus attention on China. It is impossible for me to believe that if China did not put serious and sustained pressure on this regime, China of course being what the regime depends upon for its very existence, that we wouldn't see an immediate change in twofold: One, with respect to this disaster; secondly, with respect to what the National League for Democracy and the National Coalition Government and the Union of Burma are asking for, which is not an overthrow, a violent overthrow. They are looking for negotiations, tripartite negotiations, between the military, between the National League for Democracy and between the ethnic minorities to come up with a resolution. That is an extraordinarily reasonable request and approach. There is no misunderstanding about that approach and there is certainly to misunderstanding about the brutal response of this regime to that approach.

Mr. FALEOMAVAEGA. If I could, I want to make sure I am clear for the record. Mr. Andrews, you are suggesting that we should tell China not to export any more military weapons to Burma?

Mr. ANDREWS. I think that would be an excellent start. In the same breath I would ask them to put pressure on China to allow immediate relief and unfettered access.

Mr. FALEOMAVAEGA. What about our export of weapons to Israel?

Mr. ANDREWS. Well, I am not an expert on Israel, Mr. Chairman. I think the world would be much better off if we had significantly less sales of weapons everywhere in the world.

Mr. FALEOMAVAEGA. My understanding is we happen to be the biggest exporter of military hardware and weapons in the world, somewhere around 30 billion plus.

Mr. ANDREWS. That is correct.

Mr. FALEOMAVAEGA. Here again, you raise a very valid point and question here in terms of China providing the weapons just as we are providing weapons to other countries. So it has to go both ways. If we are going to expect this of China, then we should be doing the same for others, because the perception among those who do not agree with our policy toward Israel and that terrible situation that we are finding ourselves in in Iraq right now is not very positive. And I think maybe it is an impossibility. We don't live in a perfect world. I understand that. Just as we are trying to say nonproliferation of nuclear weapons, it is a sham.

Five permanent members of the Security Council tells the world not to have nuclear weapons, but it is okay for China, France, the United Kingdom, the United States and Russia to continue having some 31,000 nuclear weapons in their arsenals. So if you were in North Korea or over in Iran, now Pakistan and India have the bomb, it is madness. And understandably providing weapons to Burma does cause a lot of serious problems over there. And the resources, but it isn't just for the resources that Burma has. But it isn't just China. There is a very strong trade relationship going on between Thailand and Burma and I believe several other countries in the Southeast Asian region.

So I just am trying to get some substance here in terms of—we talk about humanitarian, but there are a lot deeper reasons and problems that we are faced with. I am trying to unravel this, at least bring it out to the open. And I am glad that Dr. Win is here with us and saying that for 36 years the military junta has been around. And I am sorry that my friend is not here to answer the question, what do you want? Why aren't you starting a revolution? After all this is how democracy is born in some other countries. If you have a military regime and dictatorship, what do you do? You get rid of them if you have to use a force of arms. Dr. Win, would you agree to that?

Mr. WIN. Well, anyway it is what you call justified struggle, justified response in a brutal situation like this. But also we are talking about politics, you know. So we are also willing to talk with them. We are always inviting them for a talk, like Tom said, but at the same time, as you said, it is a justified choice, an alternative and somewhat starting. So we can't say eliminate that. We want a peaceful one, but with this kind of situation—but then the question is, for example, even in the United States history some country supports the United States, like France against a British colony, you know? So this is also one example I am talking about helping us because even the United States is studying states. When they started, they need the international help. We also need the international help.

Mr. FALEOMAVAEGA. My point of starting, I am glad you mentioned this. So the referendum has taken place and we approved the proposed Constitution. What is your position? Would you be willing to meet with the military government in reference to the proposed Constitution that can be implemented? Would you agree to the implementation of the proposed Constitution?

Mr. WIN. No. As it is, no. We are asking them to review it, to review it and some of the clause we may accept it, although we may not like it, but the main——

Mr. FALEOMAVAEGA. But you want to renegotiate the whole Constitution?

Mr. WIN. Well, if they are willing, but as it is we will never accept that.

Mr. FALEOMAVAEGA. So we are back to ground zero, again.

Mr. WIN. Yeah.

Mr. FALEOMAVAEGA. Congressman Andrews.

Mr. ANDREWS. Mr. Chairman, with respect to the Constitution you heard from the Deputy Assistant Secretary of State and the concerns that he had with it. But something that should also be added, during this so-called referendum on the Constitution, it was illegal to criticize the Constitution. Members of religious orders were prohibited to vote in this Constitution. So from any vantage point of an observer looking at this Constitution as an agency for democracy it clearly is not. It is a way of consolidating power for the regime and again creating the appearance of movement toward reform and progress, when in fact just the opposite is occurring.

Mr. FALEOMAVAEGA. I had a very different perception from many of my colleagues. When you talk about unification, the situation with North and South Korea, I tell my friends in Korea the only way that your people would ever become united again is that the

Koreans themselves will have to resolve that problem. The only way that Dr. Win as the Prime Minister in the elected government that they had, with that and with the military regime, you have to do this yourselves. The United States or any other government in the world will not be able to do this for you. I think this is the problem that I see sometimes. Do you want us to come do it for you?

Mr. WIN. No, no, no, not that way. We are talking about humanitarian, immediate help to those people, not to overthrow the military by the United States. That is not what we are asking. We are not asking for the regime change. That we have to do ourselves. We agree totally with you. And we will do everything we can to do that, but we want the U.S. and U.N. and our international community to keep up their strong stand, morally, politically and help us, you know, but we will do our work of course. We don't ask the United States to go in and change the regime. We don't ask——

Mr. FALEOMAVAEGA. We tried that in Iraq and look where it has gotten us.

Mr. Prime minister, I just wanted to say that, yes, partially we were talking about humanitarian aid, but the question that was raised also, why the aid has been denied or the policy of the current government said bring all the aid you want but you cannot distribute it, we will do the distribution. It goes to the more fundamental question of why? Why is it that the military regime has refused to accept this? And that is the reason why we raised the questions that are even more fundamental than humanitarian assistance. I hope you understand that.

Mr. WIN. Yeah, I agree.

Mr. FALEOMAVAEGA. I am not trying to oppose the efforts that you are making, but since my friend has labeled the current regime as the evil empire, I want to know what options are there for us if you want to get rid of the evil empire? And that is where the situation becomes a little cloudy and we are not able to provide the solution for the problems.

And we have come to the point now Burma and Darfur are not very different. Here is a crisis, literally bordering on genocide and the most powerful countries in the world cannot do anything to help. To me that is shameful. The largest international organization, the United Nations, is also helpless. And I wish there were answers in how we can resolve problems that go beyond talking about military regimes or talking about anything else where there is a moral responsibility of the world community, among the most powerful nations. My sense is why can't we just go right in there and take over? Well, you say you are destroying the sovereignty of that country for whatever it stands for. It is a nondemocratic country. So are many other nations that are not democratic. That is where the challenges lie right now and the problems we are facing. I agree with you.

Mr. Andrews.

Mr. ANDREWS. I think you are right. Clearly the future of Burma lies within the nation of Burma. And I have been moved by the extraordinary courage demonstrated by the people last fall. Those monks in the streets of Burma calling for change. The people who have lost their lives, the thousands who have lost their lives in

Burma seeking to change their country. The fact of the matter is, as you say, there are powerful forces outside of Burma that are impacting Burma. And we need to recognize the fact that if the regime is in place because of some of these powerful forces, and it seems to be incumbent upon us as a great nation to do everything that we can to try and influence those forces.

Mr. FALEOMAVAEGA. Dr. Win. Did you have anything further?

Mr. WIN. Yes, I agree with Tom that the military itself is getting help actually from China and this, what we call this constructive engagement from REM. Also we will do our job, we will do whatever we can to stop this military. The problem lies deeper, as you said. We agree. But we need help; we need help from the democratic countries, because the U.S. and others are declaring the same principle, the same value. I don't know what about United States policy to other countries, but at least with respect to Burma it is a clear cut situation where the military hold on power, negating all democratic values, negating human rights. You heard about forced labor, you heard about in Karen state this forced relocation, thousands of people. So it is very, very important that democratic countries help us whenever possible and take a strong stand, but we are not asking that United States to come and overthrow the military.

Mr. FALEOMAVAEGA. Well, I hope there was no misunderstanding of that point, but I am glad you clarified your position, Dr. Win. And thank you both, thank you Congressman, for being here. I definitely will plan to hold another hearing hopefully with more positive results on the humanitarian assistance that has been given, but I do want to thank both of you for taking the time to come and testify before the subcommittee.

As you know, I have only been chairman now a little more than a year. You will probably see my ugly face here as long as I can be in that capacity of chairman of the Asia Pacific Subcommittee. Twenty years ago nobody wanted to be on this subcommittee. The whole focus was on Europe and the Middle East. Nobody wanted to talk about Asia Pacific issues. It has only been in the last 3 or 4 years that now my own Government, my leaders are finally beginning to see the light, and hopefully see the fact that we better pay more attention to the Asian Pacific region in terms of its political, and economic and social development.

I want to wish you, Dr. Win, all the best. I am still waiting for my visa from my friends in the Myanmar, Burmese Embassy. I really would like to go there someday if they will allow me to visit. Congressman Andrews, thank you again and, Dr. Win, thank you. Hearing is adjourned.

[Whereupon, at 12:35 p.m., the subcommittee was adjourned.]

APPENDIX

Material Submitted for the Hearing Record

Prepared Statement of the Honorable Dan Burton, a Representative in Congress from the State of Indiana

I would like to thank Chairman Faleomavaega for calling today's hearing on an issue that is very important to the Asia Pacific region and to the people of the United States.

In the aftermath of the Cyclone we have seen tremendous support by the American people for those living under this brutal military junta. I would like to express my sorrow and grief for the suffering that has taken place over the past two weeks. The American people are praying for relief and we are doing everything we can to get our promised aid to those who need it the most.

It is tremendously unfortunate that, in the aftermath of a horrific natural disaster, we can once again see the evidence of a government that will not compromise its power for anything. The Burmese regime has continued to violate the freedom and well-being of its people by blocking aid to the regions of the country that are in the most desperate need. I have already heard the reports of a backlog of supplies at the Rangoon airport. I know that many countries, including the United States, have sent aid workers and supplies but are waiting in Bangkok, Thailand for the approval of their visas.

In addition, it seems to me that the decision to go ahead with a constitutional referendum that seeks to give legitimacy to the regime is a slap in the face to those who are currently suffering under it. The President and Congress expressed opposition against this referendum well before this cyclone hit. We have made efforts to weaken the junta by passing the JADE Act and refusing to purchase gems that enrich and perpetuate the influence of those who are holding onto power.

Now, in Burma's time of great need, it is our sincere desire to help the people in whatever way we can. We have already pledged millions of dollars in aid. It is my hope that we can find a way, during this hearing, to get the help to those who need it. At the same time we must continue to pressure this illegitimate regime to stop denying basic human rights to its people.

Thank you Mr. Chairman for the time, and I look forward to hearing from our distinguished panelists.

Prepared Statement of the Honorable Christopher H. Smith, a Representative in Congress from the State of New Jersey

Mr. Chairman, thank you so much for allowing me to submit a statement for the Subcommittee on Asia, the Pacific, and the Global Environment. I appreciate the great work you do here, and I appreciate in particular your calling this important hearing today. As you know, I have been active on Burmese human rights issues for many years, so I am following closely the still-unfolding tragedy of Cyclone Nargis.

I won't include here the facts and figures that illustrate the depth and seriousness of the situation—the death and destruction the Burmese people face. I know that our witnesses will outline the sad realities there.

But I do want to describe an image I saw when I was going over the heartbreaking coverage of the situation: a crowd of Burmese children, their arms outstretched, straining to try to get a handful of rice from an aid worker. You can only see the mass of humanity in the photo, but we've heard about the devastation that surrounds them—land still underwater, bodies floating amidst debris.

Looking at the faces of these children brought home to me both their pain and their courage. But the image also frustrated me, because so much more could be done right now to help these children and their families get food, medicine and drinkable water.

The United States has led the way by providing $17.5 million of humanitarian aid to help the 2.5 million people affected by this disaster—many of them in dire need of assistance.

But over two weeks after the cyclone hit Burma, the military junta is still thwarting relief efforts, refusing to grant visas to most international relief workers and restricting access to some of the victims. Some reports claim that they have even commandeered aid supplies and withheld them from the most needy.

What a tragedy that these thugs, who have denied their people so many of the basic freedoms of life, are now standing in the way of lifesaving aid!

As I said on the Floor of the House last week when supporting a resolution on this subject, this is not about scoring points on the Burma question. Our feelings on this repressive regime are well known.

This is about the children in that picture and their families.

It's the year 2008; the world has the capacity to do a great deal to mitigate suffering when a disaster like this occurs. It's a tragedy when our hands are tied by tyrants.

I'm looking forward to reading the testimony of our witnesses addressing other important issues, such as the junta's decision to move forward with its sham constitutional referendum in the midst of this catastrophe.

But at this point, it's the relief efforts and human suffering that I think preoccupy us all. Let us continue to do all we can to address these challenges.

Prepared Statement of Mr. Joel R. Charny, Vice President for Policy, Refugees International

I would first like to thank Representative Eni Faleomavaega, the Chairperson of the House Foreign Affairs Subcommittee on Asia, the Pacific, and the Global Environment, for the opportunity to submit written testimony for the May 20 hearing on Burma in the Aftermath of the Cyclone. The focus of my testimony will be on the best approach to improving the plight of the two million people affected by this catastrophe.

Cyclone Nargis was the perfect storm. It struck at the worst possible place at the worst possible time. The Burmese people, especially in the Irrawady Delta region, will be dealing with the consequences of the cyclone for many years.

Cyclone Nargis struck at the worst possible place in two senses. First, it struck the country full force at its most vulnerable spot, the low lying delta region. There was no natural barrier to impede the storm as it swept up from the coast through the country's major city, Rangoon. Second, the region is the rice bowl of Burma, and disruptions in rice production there will have negative ramifications for food availability in throughout the country, especially in the context of regional and global price increases and shortages.

The storm struck at the beginning of the rainy season, when preparations for the main rice crop were underway. Not only did the storm kill more than 100,000 people, mainly in the delta, but it swept away draft animals, destroyed dikes, and flooded fields that need to be planted by the end of June.

The timing of the cyclone also could not have been worse politically. It struck Burma exactly one week before the military government's national referendum on the new constitution, which the Burmese political opposition and ordinary citizens have dismissed as the culmination of an illegitimate process calculated to entrench the military in power. The government, suspicious of outside interference at all times, was especially concerned about externally fomented unrest in the days prior to the referendum. The senior generals, who in any event would hardly have been inclined to accept a major foreign presence overseeing the emergency response, had one more justification for placing severe limits on the international aid effort, even in the face of a disaster on the scale of Cyclone Nargis.

The cyclone will exacerbate existing vulnerabilities of the Burmese people. Under-5 child mortality is 104 per thousand, the highest rate outside Africa except for Afghanistan. HIV infection rates are the highest in Southeast Asia and malaria, a treatable and preventable disease, is the leading cause of mortality and morbidity. A rice bowl for Southeast Asia at independence in 1948, Burma is the only country in the world where Beri Beri, a vitamin deficiency, is a major cause of infant mortality; 30% of children under five are malnourished. Reflecting the difficulty that Burmese families face in feeding themselves, the average family spends 75% of its income securing adequate food supplies, compared to 57% in Cambodia and 52% in Bangladesh.

More than two weeks after the cyclone, the relief effort remains feeble. The Burmese government is primarily responsible for the frailty of the response. While it has backed off from its initial position that it could handle the emergency with its own resources, it has refused to accept international offers of aid on a scale commensurate with the need. It has allowed aid to dribble in—a few flights here and there, more visas to international personnel—but it has not made it possible for a tsunami-size effort to go forward. And the crime is that the logistical assets to undertake a major effort in the delta with helicopters and boats have been readily available in the region. But because they flew a U.S. flag, they have not been utilized. Hundreds of thousands of Burmese in the delta remain isolated and in distress.

The role of the Burmese government is inevitably problematic. The country's military leaders are out of touch with the desperate conditions of the people, as evidenced by their shock at the poverty-driven protests of the Buddhist monks last September. The authoritarian system in Burma discourages local initiative, which is critical to any emergency response. Early reports from Rangoon indicated that soldiers and police were inactive, presumably awaiting instructions on how to provide assistance from officials that had been caught by surprise by the magnitude of the disaster. In recent days, the military has been more visible, but more for show piece distributions for propaganda purposes rather than for sustained aid that would really make a difference to the survivors.

The Burmese government does not have the institutional capacity to provide relief on a massive scale. The combination of institutional weakness and suspicion of outsiders is crippling the emergency response.

In the face of Burmese intransigence, the United States, the United Kingdom, China, India, and the member states of the Association of Southeast Asian Nations (ASEAN) are coalescing around a British proposal that would have ASEAN be the face of the relief effort, using Asian disaster response experts and military assets of ASEAN countries, backed with funds, materials, and logistical support provided by the major powers. This is the only viable approach at the moment, relying on Burma's own expressed willingness to cooperate with ASEAN, but its success is far from assured due to ASEAN's lack of internal unity on matters related to Burma policy and its complete lack of experience in organizing a collective emergency response of any kind, much less on the scale required in this instance.

The core of the relief response to date has depended on self-help efforts of the Burmese people, through spontaneous efforts by concerned citizens, as well as ones by local non-governmental organizations. For the international effort, experienced relief organizations that had a presence and extensive local staff prior to the cyclone have been in the best position to respond. Ten UN agencies and 48 international NGOs were already operating inside the country with government permission. In many cases, through patient work over many years, these organizations have devised ways of operating independently of the government, mainly through their local staff working closely with Burmese community-based organizations. These staff are hired free of government interference, and they deliver assistance directly at the village level.

Given the difficult working environment in Burma, and the regime's mistrust of international actors, donor governments should rely on the capacity of organizations already inside the country as the quickest route to providing services to disaster-affected communities. Donors should ensure that NGO appeals are fully funded, and that priority is given to agencies with a proven ability to work in Burma.

I am pleased to recognize that the U.S government has adopted this approach in its initial response to the emergency. Refugees International especially appreciates the U.S. flexibility in proceeding with this funding despite the fact that its own personnel, in the form of a full disaster response team, has not been able to enter the country due to government visa restrictions.

In the medium-term, adequate response in Burma will require the presence of new international agencies. The UN and ASEAN should lead discussions with the government on streamlining procedures to register new operational agencies and managing access. In the meantime, agencies that are interested in becoming operational should explore partnerships with agencies already present, and the possibility of integrating their staff with these partners until they can set up their own official presence inside Burma.

A major medium-term challenge will be the need for recovery and development assistance. Cyclone Nargis has left several million Burmese homeless. Many villages are flattened and delta communities are reporting 90–95% damage. The threats to the 2008 rainy season rice crop and the future productivity of the delta are severe.

Rangoon, the country's largest city and economic hub, has also been directly affected. Large investments will be required to rebuild its infrastructure. This will require a long-term commitment from donors for the stabilization of the disaster-affected population and for the reconstruction of cities and villages throughout the delta, including Rangoon.

Currently, most donor nations have strict restrictions on the provision of development assistance to Burma, as this type of aid is usually provided for cooperative projects with the government. These restrictions are useful insofar as they ensure money is not misused by the Burmese regime. The demands for reconstruction aid will be substantial, however, and the United Nations, in cooperation with international NGOs, will need to define how best to carry out this work while ensuring the greatest degree of independence possible.

At the same time, members of Congress should begin to develop policy options that allow for development-style assistance to Burma within politically acceptable limits. It should begin to do this in consultation with NGOs working in the country to ensure that political limitations and operational needs complement each other, as is the case with the current European Commission Common Position on Burma.

It will also be important to extend programs beyond the disaster-affected areas to the country as a whole. The loss of food supplies and farmland in the delta region, the nation's rice bowl, could have negative consequences for highly vulnerable people in other parts of the country. Similarly, the further sapping of Rangoon's economic strength in an anemic economy could have reverberations throughout the country that will further jeopardize livelihoods in areas that were not directly affected by Cyclone Nargis.

The political impact of the cyclone is impossible to predict. The differing post-tsunami experiences of Indonesia and Sri Lanka point to the difficulty of judging the cyclone's ramifications in Burma. In Aceh, the severity of the tsunami broke the political impasse between the armed resistance and the Indonesian government, freeing both parties from long-held rigid positions as they gradually coalesced in the interests of the welfare of the people. A dramatically increased international presence helped create the environment for these developments. In Sri Lanka, in contrast, the tsunami response quickly became politicized, amid mutual accusations of unjust aid allocations and donor bias, which contributed to the return to open warfare between the Tamil Tigers and the Sri Lankan government.

The cyclone offers the possibility of revitalizing the relationship of the Burmese government and the international community as the world's generosity manifests itself in the coming days and weeks. The hope is that the scale of the disaster is so immense that even the reclusive military leaders, at ASEAN's urging, will eventually have no choice but to accept a large-scale international aid presence. For mid-level civilian officials, the new engagement with the outside world will be a welcome opportunity. Even if the generals who run Burma make it difficult for the aid agencies to respond to needs in keeping with humanitarian principles and practice, new relationships will be forged at the local level that will bring a measure of hope to the long-suffering Burmese people.

The American people traditionally show strong support for assistance to those in dire need, regardless of their nationality, religion or form of government. After Hurricane Mitch, the tsunami in Southeast Asia in 2004 and the earthquake in Pakistan in 2005, Congress passed supplemental spending bills that authorized multi-year commitments of funds for the emergency response, and we feel a similar commitment is needed for Burma.

The financial requirements for the emergency response and the near-term reconstruction effort will amount to more than two hundred million dollars. At this point the United States is largely doing the right thing—stressing the humanitarian needs and the imperative to respond; making generous offers of assistance; supporting the diplomatic efforts of the UN Secretary-General, ASEAN, and regional powers with the Burmese government. In closing, I urge Congress to give the Administration the financial resources that it needs—$40 million—through the supplemental appropriations legislation currently under consideration to ensure that the United States is able to play an appropriately strong role in the response to the Cyclone Nargis catastrophe.

Testimony of Gregory Beck
Asia Regional Director
International Rescue Committee

Before the
COMMITTEE ON FOREIGN AFFAIRS
SUBCOMMITTEE ON ASIA, THE PACIFIC, AND THE GLOBAL ENVIRONMENT
Hearing on
"Burma in the Aftermath of Cyclone Nargis:
Death, Displacement, and Humanitarian Aid"

May 20, 2008
Submitted for the Record

Thank you for holding a hearing on Cyclone Nargis and its impact. The International Rescue Committee is submitting this short testimony for the record in order to provide information to the subcommittee from the vantage point of an aid agency operating on the ground. I can report that IRC emergency personnel are succeeding in delivering life-saving health and shelter supplies to cyclone victims as are other aid organizations, but the amount of assistance reaching survivors does not come close to what is needed to address a crisis of this scale. Access to impacted communities remains extremely limited and I would like to share this and other concerns in this testimony.

Dimensions of the Catastrophe

Cyclone Nargis pummeled Myanmar over two weeks ago, on May 2 and 3. The low-lying Irrawaddy delta region was devastated by the cyclone. The official death toll now stands at 78,000 with 56,000 missing. Unofficial estimates are considerably higher. More than half a million people are estimated to have gathered in improvised camps scattered across the delta region. The UN now estimates that as many as 2.5 million people have been severely affected.

The government has declared 45 townships *disaster areas* and formally requested support from the international relief community. The cyclone devastated much of what was in its path, flattening and flooding villages, knocking out power lines, polluting fresh water sources and farm land with salt water, blocking roads and destroying infrastructure.

Many of the homeless are migrating away from the worst affected areas towards higher and drier ground or the larger towns. The government has established some camps in the main townships, but many more makeshift settlements have formed in monasteries, schools, empty buildings or open spaces. Some receive assistance from local authorities and monasteries and many have received no aid at all. More and more local groups and citizen volunteers are out on the roads distributing supplies and caring for the desperate.

IRC'S Response

A small number of aid agencies had a presence in Myanmar before the cyclone. The IRC was among the aid groups that rushed to the scene in its wake, with a team of international IRC emergency specialists dispatched to the region within days of the cyclone hitting. So far, only one has been able to obtain a business visa to enter the country. The rest are providing intensive technical support from Thailand, while working round the clock to procure emergency supplies.

In spite of the hurdles, the IRC decided not to wait for additional staff and supplies to arrive. The IRC's emergency coordinator, Gordon Bacon, was able to conduct a rapid assessment upon arrival in Myanmar, assemble a team of local staff and volunteers, forge a partnership with a local aid agency, identify a local supplier and two trucks, and begin emergency distributions to a cyclone-ravaged district on the southeastern edge of the Irrawaddy Delta.

Over three days last week, the team delivered critical emergency supplies to hundreds of families, some 1,500 people, who had yet to receive any aid in Kunyangon Township. The materials included all locally-purchased mosquito nets, blankets, tarps, clothing, water containers, cooking supplies and candles. In one settlement, there were 300 survivors who had lost their homes and were clustered in a structure that gave them no protection at all because the roof was torn off by the cyclone.

Gordon Bacon has reported to us about the sheer misery and desperation these people are suffering. "They wait for help – homeless, cold, sick and hungry – and get lashed by more rain. So the sick get sicker when they need to stay dry and warm and get proper care. We are providing blankets and tarps in order to give some measure of protection from the elements. Frankly, it's modest assistance compared to what is needed in a crisis of this magnitude," Gordan reported.

IRC staff and volunteers also conducted an assessment in Wakema Township just north of the delta – a district that was lashed by the cyclone, flooding 25 villages, and is now struggling to absorb thousands of homeless survivors from other impacted areas. A first IRC distribution was conducted here yesterday, in advance of planned health, sanitation and shelter programs.

The IRC is also poised to receive 40 tons of donated and purchased medical, water treatment and shelter supplies. The emergency stocks are arriving in Myanmar on international flights, the first being a generous USAID shipment of plastic sheeting, hygiene kits and water containers that arrived Saturday and is being readied for distribution in conjunction with local partners. Additional contributions of supplies are coming from the Spanish Government (supplies for prevention of waterborne diseases) and Muslim Aid and Global Medics (medicine and water purification equipment). The IRC is also sending pre-positioned emergency supplies from its warehouse in Dubai. In all, the supplies will benefit an estimated 80,000 people.

As always, we are also seeking financial support from varied donors, including foundations and the concerned public.

Improvements, Challenges and Concerns

This week has already brought some encouraging news. While monsoon rains are making aid deliveries all the more difficult, "a new cyclone" that had been forecasted to hit Myanmar did not come to pass. In the meantime, the number of flights arriving in Yangon with crucial supplies has picked up and aid is gradually reaching devastated communities. Myanmar also agreed to receive assistance from the Association of South-East Asian Nations (Asean) and invited a cadre of Asian medics from India, Thailand, China and elsewhere to provide health assistance.

Nevertheless, we remain extremely concerned that this external assistance will come too late for thousands of people who now suffer from malnutrition and disease, conditions that could have been preventable had the international aid community been granted full access to cyclone survivors in a timely manner.

The disaster in Myanmar requires a tsunami-like international relief effort. In that emergency response, a massive influx of international health professionals arrived quickly and joined forces with local expertise to treat the wounded, provide clean drinking water and sanitation, and prevent outbreaks of contagious diseases.

The exact opposite has taken place in Myanmar. Two weeks into the disaster, the UN provided the sobering statistic that little more than 20% of people gravely impacted by the crisis are receiving aid.

Most survivors have no access to clean drinking water, as virtually all water sources have been contaminated by human waste and decaying animal carcasses, and containers to collect rain water were washed away. It takes less than a spoonful of contaminated water for a person to come down with dysentery or cholera -- highly infectious diseases that can quickly kill in the tens of thousands. It's no wonder that reports of both disease are on the rise. The threat of a second wave of deaths from a public health crisis is upon us.

The pace of the response has not been fast enough and not enough aid has arrived in the country. Aid to date is far below what is required to begin to meet the needs in an emergency of this scale. Aid should be allowed to reach areas beyond Yangon and we would like to see more international relief workers let into the country.

In light of the magnitude of this crisis, a massive infusion of aid and experienced disaster response experts is needed to prevent a public health catastrophe. We ask that the Congress include increased aid to the cyclone's survivors as one part of the supplemental appropriations bill now under consideration. We are well aware of competing needs – the IRC runs humanitarian aid programs in 25 conflict zones, including Iraq and Afghanistan, and we are also asking you to address problems caused by the global increase in food prices and the erosion in funding for humanitarian needs in neglected crisis areas such as the horn of Africa, We ask for a great deal, but only because many, many lives are at stake and the United States can make a difference and save tens of thousands of lives with little effort.

In conclusion, the IRC and other operational aid groups are getting supplies and materials directly and safely to the cyclone's victims and saving lives. We and other aid groups can do much more if sufficient funding is provided and emergency professionals have the access and equipment they need to help people quickly and efficiently. We appeal to you for help. Time is of the essence.

Thank you very much.

———

USAID
FROM THE AMERICAN PEOPLE

CYCLONE NARGIS

USAID Responds to Cyclone Nargis

KEY DEVELOPMENTS

July 30, 2008

Photo Gallery: USAID Administrator Delivers Aid to Burma and Announces an Additional $13 Million in Humanitarian Assistance

- On July 21, the Tripartite Core Group, including representatives from the Association of Southeast Asian Nations (ASEAN), the Government of Burma (GOB), and the U.N., released the final Post-Nargis Joint Assessment (PONJA) report, based on assessments conducted in Ayeyarwady and Rangoon divisions from June 10 to 19. According to ASEAN, the document describes the human loss and damage to infrastructure caused by the cyclone, as well as the impact of the disaster on both the national economy and household-level livelihoods.
- On July 10, the U.N. released a revised $482 million appeal to support post-cyclone assistance for Burma through April 2009. The appeal, based on the PONJA findings, includes $201 million requested in the previous U.N. appeal.
- USAID/OFDA is supporting additional agricultural and livelihoods recovery activities through the international non-governmental organization (NGO) Save the Children/U.S. (SC/US). USAID/OFDA is also supporting the U.N. Development Program (UNDP) and the NGO PACT to recapitalize an existing microfinance program disrupted by the cyclone.

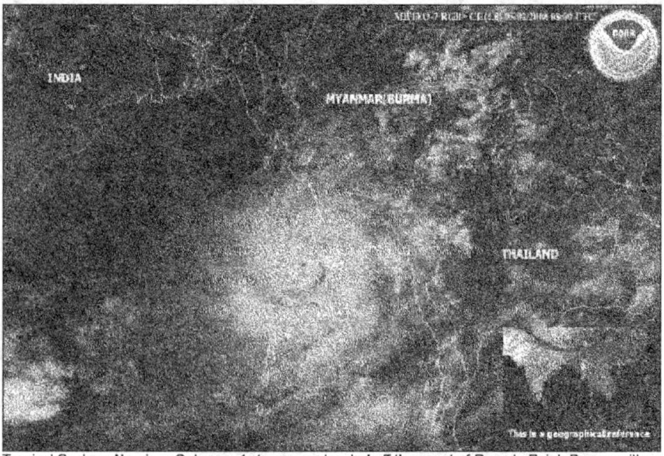

Tropical Cyclone Nargis, a Category 4 storm, was located off the coast of Pagoda Point, Burma, with sustained winds of 130 mph and gusts up to 160 mph.

ESTIMATED NUMBERS AT A GLANCE*		SOURCE
Total Dead	84,537	GOB[1] – June 24, 2008
	63,000-101,000	OCHA[2] – May 9, 2008
Total Missing	53,836	GOB – June 24, 2008
Total Number Affected	2.4 million	OCHA – May 29, 2008

FY 2008 HUMANITARIAN FUNDING PROVIDED TO DATE

USAID/OFDA Assistance to Burma: $28,506,180
USAID/FFP[3] Assistance to Burma: $12,000,000
DOD[4] Assistance to Burma: $9,475,900
Total USG Humanitarian Assistance to Burma: $49,982,080

1. U.N. Office for the Coordination of Humanitarian Affairs (OCHA)
2. USAID's Office of Food for Peace (USAID/FFP)
3. U.S. Department of Defense (DOD)

CURRENT SITUATION

- According to OCHA, the cyclone severely affected an estimated 2.4 million people. During a three-day visit to Burma that concluded on July 24, the U.N. Under-Secretary-General for Humanitarian Affairs indicated that humanitarian organizations had reached virtually all cyclone-affected individuals with some relief assistance, but stressed the challenge of systematically providing ongoing support, particularly to populations in remote areas.

Early Recovery

- From July 8 to 11, a USAID/OFDA disaster specialist, a USAID Regional Development Mission for Asia medical doctor, and three staff from the U.S. Embassy in Rangoon conducted general needs assessments in

11 villages in Labutta and Ngapudaw townships, Ayeyarwady Division. Residents surveyed by the assessment team in each village consistently identified assistance to reestablish livelihoods as a primary need.

- USAID/OFDA is focusing on livelihoods interventions to provide quick access to income and employment for affected populations. USAID/OFDA is supporting NGO partners including Church World Services (CWS), International Development Enterprises (IDE), SC/US, World Concern (WC), and ACTED to implement early recovery activities for cyclone-affected populations in Ayeyarwady and Rangoon divisions. The programs will assist vulnerable communities to resume agricultural activities during the current planting season, as well as providing employment opportunities, supporting the resumption of interrupted livelihoods, recapitalizing a microfinance program, and rehabilitating basic infrastructure in affected communities.

Food Assistance

- On July 25, the U.N World Food Program (WFP) noted that the PONJA findings substantiated WFP's concerns regarding a drastic reduction in household food stocks as a result of the cyclone. The PONJA indicated that more than 40 percent of the cyclone-affected households surveyed lost all food stocks during the cyclone.
- In response to the PONJA results, WFP recently scaled up emergency feeding programs for 924,000 beneficiaries, which will continue until April 2009 at a total cost of $112 million. To date, WFP has assisted 733,490 cyclone-affected individuals with food assistance, in addition to coordinating the U.N. Logistics Cluster.
- To date, USAID/FFP has provided approximately $12 million in P.L. 480 Title II emergency food assistance through WFP. USAID/FFP is planning to make an additional contribution to WFP to support expanded WFP operations. USAID/FFP will continue to monitor food security in cyclone-affected areas.

Logistics

- On July 28, the U.N. Logistics Cluster noted that the air bridge from Bangkok to Rangoon now comprises two Antonov-12 aircraft, one of which has replaced the larger capacity Ilyushin-76 aircraft. The U.N. logistics cluster has reported that the air bridge will conclude operations on August 10.
- The U.N. Logistics Cluster in Rangoon conducted a workshop on July 28 to plan for the continued operation of the logistics hubs in the Ayeyarwady Division by humanitarian NGOs after August 10. Participating organizations discussed the existing parameters of hub operations, addressing possible cost-sharing options between organizations and identifying potential lead agencies for the hubs of Pyapon, Bogale, and Labutta townships.
- The cluster reported that boats and barges are now transporting as much as 95 percent of the surface cargo to the logistics hubs due to the continued deterioration of roads as the rainy season progresses.

USG HUMANITARIAN ASSISTANCE

- On May 5, U.S. Chargé d'Affaires Shari Villarosa declared a disaster in Burma due to the effects of Cyclone Nargis. In response, USAID/OFDA deployed a USAID Disaster Assistance Response Team and immediately provided $250,000 to U.N. agencies for shelter and water and sanitation assistance.
- To date, USAID/OFDA has provided a total of more than $28 million for emergency relief commodities and programs in agriculture and food security, economy and market systems, health, nutrition, protection, shelter and settlements, humanitarian coordination, logistics, and water, sanitation, and hygiene.
- From May 12 to June 22, the DOD-operated USG air bridge completed 185 airlifts and delivered more than $4 million in USAID/OFDA emergency relief supplies, as well as DOD, U.N., NGO, and Government of Thailand commodities. The USAID/OFDA commodities benefited at least 445,000 individuals and included 38,800 hygiene kits, 6,500 rolls of plastic sheeting, 15,000 mosquito nets, 95,300 ten-liter water containers, 10 water treatment units, 12 zodiac boats, and fourteen 10,000-liter water storage bladders. USAID/OFDA funded 12 flights from regional warehouses in Dubai and Pisa to deliver the commodities to Utapao, Thailand, for transportation to Burma on the USG air bridge.

Bloated corpses pile up; first U.S. aid flight arrives in Myanmar

- Story Highlights
- **NEW:** U.N. chief blasts Myanmar junta's response to aid offers
- Bodies taint Myanmar water supplies
- U.S. plane takes relief supplies; clearance granted for two more flights
- U.N. estimates death toll from Cyclone Nargis is 63,000 to 100,000

YANGON, Myanmar (CNN) -- Bodies continued to pile up in Myanmar as the first U.S. aircraft carrying relief supplies touched down Monday in the cyclone-ravaged country.

The shipment arrived in a time of dire need in the country's poorest regions, where people have been reduced to pumping water from ponds filled with the dead.

The first airlift, a C-130 Hercules loaded with 28,000 pounds of supplies, including water, mosquito netting and blankets, landed in Myanmar about 2 p.m.

Two more U.S. Air Force planes are expected to arrive in Myanmar Tuesday with more humanitarian supplies. White House spokeswoman Dana Perino also announced Monday that the United States has offered Myanmar an additional $13 million in aid, bringing the total to $16.25 million.

Though the Myanmar government recently began allowing international aid to enter the country, it was unclear when that aid would reach villages in southern Myanmar.

The country's ruling military junta is distrustful of Western countries.

A CNN correspondent who is not being identified for his safety said he saw the country's militia delivering international aid in some bigger cities. But some southern villages seemed overlooked.

"They have no drinking water whatsoever," the correspondent said. "When you don't have drinking water and you are forced to drink out of puddles and drinking reservoirs contaminated by dead bodies... It is a very dire situation."
Watch how rotting corpses line the riverbanks »

More than a week after the cyclone hit the south Asian country, getting relief there has been a daunting task for international aid agencies.

The Britain-based international aid agency Oxfam warns that without the proper relief -- particularly clean water -- nearly 1.5 million people could be affected by a wider humanitarian crisis. Watch a report on widespread death and destruction »

A refugee camp in Pyanpon township was operating with five latrines for 3,500 people, UNICEF said.

The shore along the Irrawaddy River Delta remains lined for miles with bloated corpses. In the village of Da Mya Kyaung, only four of the 200 homes were partially intact.

"When I saw the water coming, I just put my two nephews on my shoulders and ran," villager U Wen Say said.

His son and his son's family drowned. Of the 500 people who lived in the village, two-thirds were missing.

The United Nations estimates the death toll from Cyclone Nargis ranges from 63,000 to 100,000, well above the Myanmar government's estimate of about 28,000. Tens of thousands of people are missing. Watch survivors await relief supplies »

U.S. officials hope the relief flights that have been approved by the Myanmar junta will forge a relationship that will allow the United States to send in disaster experts.

"As of right now, visas for them have not been approved," White House spokesman Gordon Johndroe said. "So we'll keep on working on this. We hope this is the beginning of a long line of assistance from the United States to the people of Burma."

The military junta has said it will accept international aid but insisted it would distribute the supplies itself.
Watch relief supplies trickle in to Myanmar »

U.N. Secretary-General Ban Ki-moon on Monday blasted Myanmar's government for impeding aid efforts, calling on the junta to "put its people's lives first."

"This is not about politics. It is about saving people's lives. There is absolutely no more time to lose," Ban said.

Four U.S. Navy ships that are in the region for an annual military exercise can also help in the relief mission if the Myanmar government gives the go-ahead.

The United States does not recognize the military junta.

Washington has been a vocal critic of the junta, which maintained control of the country even after 1990, when an opposition political party won victory in democratic elections.

Debbie Stothard, head of the Southeast Asian human rights group ALTSEAN-Burma, said her organization has received reports of aid packages being distributed with the names of military leaders on the labels.

"There's people who are very concerned now that the reason the aid workers are being blocked is so that the military can deliver aid selectively and so that they can appropriate the aid and pretend it was from them in the first place," Stothard said.

The country's name was changed from Burma to Myanmar in 1989, but many who do not recognize the current government still use its former name.

All AboutMyanmar • United Nations World Food Programme

Links referenced within this article

Myanmar
http://topics.edition.cnn.com/topics/Myanmar
Watch how rotting corpses line the riverbanks »
#cnnSTCVideo
Watch a report on widespread death and destruction »
#cnnSTCVideo
Watch survivors await relief supplies »
#cnnSTCVideo
Watch relief supplies trickle in to Myanmar »
#cnnSTCVideo
Myanmar
http://topics.edition.cnn.com/topics/Myanmar
United Nations World Food Programme
http://topics.edition.cnn.com/topics/United_Nations_World_Food_Programme

Find this article at:
http://edition.cnn.com/2008/WORLD/asiapcf/05/12/myanmar.aid/index.html

Uncheck the box to remove the list of links referenced in the article.

newsVOA.com

Print

US Raises Burma Relief Donations to $16 Million

By Luis Ramirez
Bangkok, Thailand
12 May 2008

Ramirez report - *Download (MP3)*
Ramirez report - *Listen (MP3)*
P Sisco Burma report / Broadband - *Download (WM)*
P Sisco Burma report / Broadband - *Watch (WM)*
P Sisco Burma report / Dialup - *Download (WM)*
P Sisco Burma report / Dialup - *Watch (WM)*

The United States has delivered its first load of humanitarian supplies for victims of Cyclone Nargis in Burma, and announced it is raising its relief donation to more than $16 million. A U.S. Air Force C-130 cargo aircraft landed in Burma's main city, Rangoon, in what Washington hopes is the start of a larger effort to help cyclone victims. The Burmese government now says more than 28,400 are confirmed dead and more than 33,000 remain missing. VOA Southeast Asia Correspondent Luis Ramirez reports from Bangkok.

The U.S. Air Force C-130 cargo aircraft left a Thai air force base in the town of Utapao loaded with pallets of bottled water, mosquito netting, and blankets - the first shipment that Burma's military government has allowed the United States to deliver.

US military personnel unload aid shipment to Burma, 12 May 2008

Burma's government had been reluctant to accept any U.S. aid, even as thousands, perhaps millions, faced the threat of hunger and disease in the aftermath of the disaster. U.S. officials called Burma's acceptance of the first planeload of supplies an important step in getting the country to start receiving more aid that survivors desperately need.

On the first relief flight was the top commander of U.S. forces in the Pacific, Admiral Timothy Keating, who met with Burmese officials in Rangoon Monday. Upon his return to Bangkok, the admiral told reporters he had offered the full help of the U.S. military. He said 4,000 U.S. Marines in the region, along with a fleet of C-130s, and helicopters which are part of joint exercises in Thailand, are prepared to help now.

"They are in place in Thailand ready to go forward as soon as the Burmese would give us permission to do so," Admiral Keating said. "We have three ships that are approaching the southwest Burmese coast in international water very close to the shoreline in 36 to 48 hours. So, we have a broad array of personnel and equipment and we're ready to respond as soon as the Burmese give us permission. We did not get that permission today."

U.S. officials describe their meetings with Burmese deputy ministers and other officials as cordial, but indicated the Burmese made few concessions. Henrietta Fore, the head of the U.S.

Agency for International Development, USAID, - also on Monday's relief flight - says progress was made.

The remains of a home in Laputta township in the Irrawaddy Delta region

"We left Utapao with the hope that we could lay the groundwork for a broader United States assistance effort and I believe that our discussions were a good first step,' she said. "We see it as a beginning, but this will take time."

Henrietta Fore greets Burmese official, as US aid shipment arrives, 12 May 2008

Fore announced the United States is raising its contributions to relief efforts in Burma from $3.2 million to more than $16 million. She said the additional $13 million will go to the United Nations World Food Program for hunger relief efforts in Burma.

The announcement came as WFP officials said they have been able to meet only 20 percent of the needs of the storm survivors who are going hungry. At the same time, the U.N.'s Food and Agriculture Organization said Monday the price of rice has gone up 50 percent in Burma's markets since the cyclone hit.

Cyclone damage in Laputta township in Irrawaddy Delta region

International aid agencies are keeping up their calls for the Burmese government to issue visas to relief workers and allow more aid to go in. They warn deteriorating sanitary conditions and the spread of disease could kill millions, especially as forecasts this week call for heavy rains in hard-hit regions of the Irawaddy Delta.

Burma's reclusive military government has been accepting small quantities of aid, but has been reluctant to admit large numbers of western relief workers.

Print

U.S. Rejects British Genocide vs. Myanmar

by Mike Billington

After a week of wild lies from Western govenments and press claiming that the Myanmar military regime was refusing emergency aid for the millions of victims of the May 3 Cyclone Nargis, and escalating calls for a military invasion for "humanitarian reasons," the head of the U.S. Pacific Command, Adm. Timothy Keating, flew into Yangon on May 12 with a delivery of food and other supplies. Keating met with the head of the Myanmar Navy, turned over the entire shipment to the Myanmar military for distribution, and assured them that the hysteria, the lies, and the threats against Myanmar were not U.S. policy.

The Admiral told NPR News May 14 that the U.S. aid was unconditional, to be distributed by the sovereign government of Myanmar, and categorically rejected the calls for an invasion. Asked by NPR if such an invasion were a "remote possibility," Keating replied, "It is not. That's why I and my State Department colleagues went to Burma. The spigots are opening. We have absolutely no intention of forcefully providing relief supplies." Asked if Myanmar were not refusing aid, as reported in news media accounts across the West, he countered that other nations, the United Nations, and many NGOs (Non-Government Organizations), were delivering extensive aid, and that with the United States now contributing, "it's having an effect." Keating said that the Myanmar government had agreed to the U.S. aid "in an interesting way—we flew in a couple of C-130s today, and by approving our flight plans, they are giving permission. We have 170,000 pounds in so far, and another five flights tomorrow, and perhaps some helicopters as well."

This is an extraordinary shift in U.S. policy. The stated policy of the Bush Administration was that the U.S. and Europe would deny any assistance to the victims of the cyclone in Myanmar, unless U.S. military teams were allowed free access to assess the scope of the crisis, followed by U.S. and other Western teams who would administer the distribution of aid. This is in keeping with Henry Kissinger's National Security Memorandum 200, signed into policy by President Gerald Ford in 1974, asserting that food should be used as a weapon to reduce population growth, and denied to nations which refuse to accept neocolonial demands on their sovereign rights. While Myanmar opened its doors to its friendly neighbors, it rejected absolutely any conditions on Western aid.

In an equally extraordinary shift, the U.S. C-130 Hercules which carried Keating and the U.S. supplies into Yangon, was shown prominantly in eight photos in the official government newspaper, *The New Light*, clearly showing the "U.S. Air Force" insignia, sitting on the Yangon airport tarmac. This meeting is by far the highest level contact between the U.S. and Myanmar in many years.

Why Myanmar Was Targetted

Myanmar has been treated as an "outpost of tyranny" (to use Secretary of State Condoleezza Rice's term), and subjected to threats and sanctions continuously for the past 20 years. The publicly stated excuse for this subversion is the so-called "human rights" abuses against British asset Aung San Suu Kyi and her opposition party within Myanmar. Suu Kyi was trained at Cambridge and married Michael Aris, one of the leading British intelligence officials responsible for the Himalayas and the Subcontinent. She has functioned as a British asset since her return to Myanmar in 1988, preventing every effort at cooperation between opposition leaders and the government, even by those in her own party who wanted to work with the government and Myanmar's neighbors, on the development of their country. The government refers to Suu Kyi as an "axe handle" for the British axe, which threatens the survival of the nation.

But the real purpose of the sanctions and the demonization of Myanmar by the British Empire and its foolish supporters in the U.S.A., is that Myanmar is the strategic hub, connecting China, India, and the Southeast Asian nations. The British granted independence to Myanmar (known as Burma, while under British colonial domination) in 1948, at about the same time they were forced to grant India its independence, but left the nation with an impossible constitution, granting the right to secession to all the ethnic groups along the border, and then sponsoring insurgencies by several of them, mostly financed by the opium trade, also sponsored by the British and their Hong Kong banks.

The military government which took power in 1988 moved successfully in the 1990s to make peace with all the ethnic insurgent armies, and nearly eliminated the opium production in the process. This revived the possibility for the first time, since the British colonization in the 19th Century, for Myanmar's cooperation with its neighbors, opening up road and rail connections facilitating regional trade and development. This development was the enemy in the eyes of British geopolitics—alliances among sovereign nations which could threaten the power of the British banking cartels and the Anglo-American control of trade through sea power.

But the attempted Anglo-American isolation of Myanmar through sanctions and threats has been a failure, since all—repeat, *all*—of Myanmar's Asian neighbors have rejected the geopolitical isolation of the country. Despite the extreme difficulties imposed by the sanctions, including the total cutoff of assistance from the Asian Development Bank and the World Bank, great infrastructure projects are well underway, linking Myanmar by road and rail to India, China, and Thailand, de-

DoD/Sgt. Andres Alcaraz, USMC

While the corrupt Western news media has been replete with stories about Myanmar's rejection of aid from the West, this photo of a U.S. Air Force C-130 Hercules aircraft delivering supplies at Yangon International Airport shows them to be outright lies. The Myanmar government has, however, absolutely rejected any British-style conditions on the aid.

veloping ports, airports, dams, oil pipelines, and other infrastructure projects. The "Southern branch" of the Eurasian Land-Bridge, connecting China with Europe and Africa through Southeast Asia, India, and Southwest Asia, is now a near-term possibility.

The Blair Doctrine

The shift in U.S. policy indicated by Admiral Keating's visit was presaged by Secretary of Defense Robert Gates' firm rejection of proposals by U.S. aid officials that the United States begin food distributions within Myanmar without government approval. This variation on neoconservative "preemptive invasion" was first suggested by French Foreign Minister Bernard Kouchner, who argued that the "responsibility to protect" clause, adopted at the 2005 UN World Summit, allowed for "coercive humanitarian aid" to be delivered to the cyclone victims, based on the lie that the Myanmar government was refusing to allow delivery of emergency supplies.

Kouchner was evoking what has become known as the "Blair Doctrine," after the open call by former British Prime Minister Tony Blair to do away with the notion of sovereign nation-states, which had been established by the Peace of Westphalia in 1648, in favor of preemptive wars in disregard of sovereignty. Blair, in the Fabian imperial tradition of Lord Bertrand Russell and H.G. Wells, presented his doctrine in a speech in Sedgefield, England on March 5, 2004, specifically defending the British and American preemptive war on Iraq, but extending it to a new global imperial order:

"So, for me, before Sept. 11, I was already reaching for a different philosophy in international relations from a traditional one that has held sway since the Treaty of Westphalia in 1648; namely, that a country's internal affairs are for it, and you don't interfere unless it threatens you, or breaches a treaty, or triggers an obligation of alliance....

"It may well be that under international law as presently constituted, a regime can systematically brutalize and oppress its people and there is nothing anyone can do, when dialogue, diplomacy, and even sanctions fail.... This may be the law, but should it be?... [W]e surely have a responsibility to act when a nation's people are subjected to a regime such as Saddam's. Otherwise, we are powerless to fight the aggression and injustice which over time puts at risk our security and way of life.

"Which brings us to how you make the rules and how you decide what is right or wrong in enforcing them. The UN Universal Declaration on Human Rights is a fine document. But it is strange [that] the United Nations is so reluctant to enforce them. But our worry is that if the UN—because of a political disagreement in its Councils—is paralyzed, then a threat we believe is real will go unchallenged. Britain's role is try to find a way through this: to construct a consensus behind a broad agenda of justice and security and *means of enforcing it*" (emphasis added).

Calls for Invasion

Joining the chorus of support for the Blair Doctrine, demanding an invasion of Myanmar, were, among others:

- British Fabian Simon Jenkins, who called for a full-scale invasion in the London *Guardian* on May 15, complains that people were dying as "our macho invaders sit on their hands.... Where are the buccaneers of Bosnia, the crusaders of Kosovo, the bravehearts who rescued Sierra Leone from its rebels, the Afghans from the Taliban and the Iraqis from Saddam Hussein?" To explain away the successful visit of Admiral Keating, Jenkins simply lies that "he was sent packing" by the junta leaders.
- EU foreign policy chief Javier Solana told an emergency meeting of EU ministers in Brussels that, "The United Nations charter opens some avenues if things cannot be resolved in order to get the humanitarian aid to arrive," threatening to use UN forces to do "whatever is necessary to help the people who are suffering."
- Gareth Evans, the former Australian foreign minister,

who now heads the International Crisis Group, fully backed French Foreign Minister Kouchner's criminal threat, bragging that he had been one of the authors of the "responsibility to protect" clause. Evans, also writing in the *Guardian*, said that, although the clause was intended to stop genocide by criminal regimes, the Myanmar regime was "denying relief to hundreds of thousands of people at real and immediate threat of death," justifying an invasion.

• Ivo Daalder of the Brookings Institution, a leading foreign policy advisor to Barack Obama, wrote that the UN must "demand that the Burmese government accept the offers of international relief supplies and personnel, without interference, and allow the UN to take charge of the humanitarian mission," or face coercive action.

• *Time* magazine of May 10 published an article titled: "Is it Time to Invade Burma?" answering in the affirmative. "If we let them get away with murder," writes author Romesh Ratnesar, "we may set a very dangerous precedent."

• Jan Egeland, former UN emergency relief coordinator, accused Myanmar's government of "murder."

• Shawn Crispin, a journalist for various Dow Jones publications in Asia, and a graduate of Johns Hopkins School of Advanced International Studies (SAIS), where neocon warmonger Paul Wolfowitz once lectured on his views for imposing "democracy" through military means, sounded very much like Wolfowitz before the Iraq invasion, in an article for *Asia Times*. Crispin argued that the Myanmar population would "warmly welcome a U.S.-led humanitarian intervention," and that the military would "defect en masse rather than confront U.S. troops." Bush could "burnish his foreigh policy legacy," Crispin proposed, by using such a preemptive war "for the good" against Myanmar.

The Reality

None of those arguing for war could have been unaware of the reality on the ground, despite massive lying in all the Western news media, but rather chose to ignore the extensive evidence provided by competent sources who were, in fact, actively engaged in the humanitarian effort within Myanmar. For example, the Red Cross has been publicly reporting on the extensive aid reaching the victims of the cyclone, despite horrendous logistical problems due to the collapse of the feeble infrastructure that existed before the storm. Red Cross spokesman Joe Lowry told Bloomberg on May 10, one week after the cyclone, that 11 planeloads of Red Cross supplies were in the country, or on the way, in addition to the aid from Myanmar's neighbors. He said that the Myanmar Red Cross had mobilized "thousands of volunteers" to help in the distribution. "I don't want to say that we haven't had difficulties," he said, "but we don't do our negotiations in public. I think we've been helped by being a neutral organization with no agenda except providing aid."

World Vision Australia head Tim Costello, who was in Yangon within a few days of the storm, told Australia's ABC News that, "The impression that no aid is getting through is wrong. We are actually getting aid to some of the most farflung areas."

The World Health Organization (WHO) told the *New York Times* May 14 that its medical supplies were arriving in the country normally, without being diverted or siphoned off, and that deliveries were reaching the hardest hit locations. There were no reports of outbreaks of malaria or dengue fever as of yet, although this remains a great danger.

Refugees International head Joel Charney told NPR News on May 6, three days after the cyclone: "There are ten UN agencies working in Myanmar, and 48 relief and humanitarian groups already in place. Outsiders underestimate the number of agencies there, and the scope of their programs. There is international work going on now in almost all of the country."

In other words, the world can survive without the Anglo-Americans running things. On the other hand, the U.S. helicopters and other military capacities would obviously be of great assistance—and now that the saner elements in Washington have rejected the British-colonial "regime change" rhetoric, perhaps they can begin to be of help.

USAID HUMANITARIAN ASSISTANCE TO BURMA

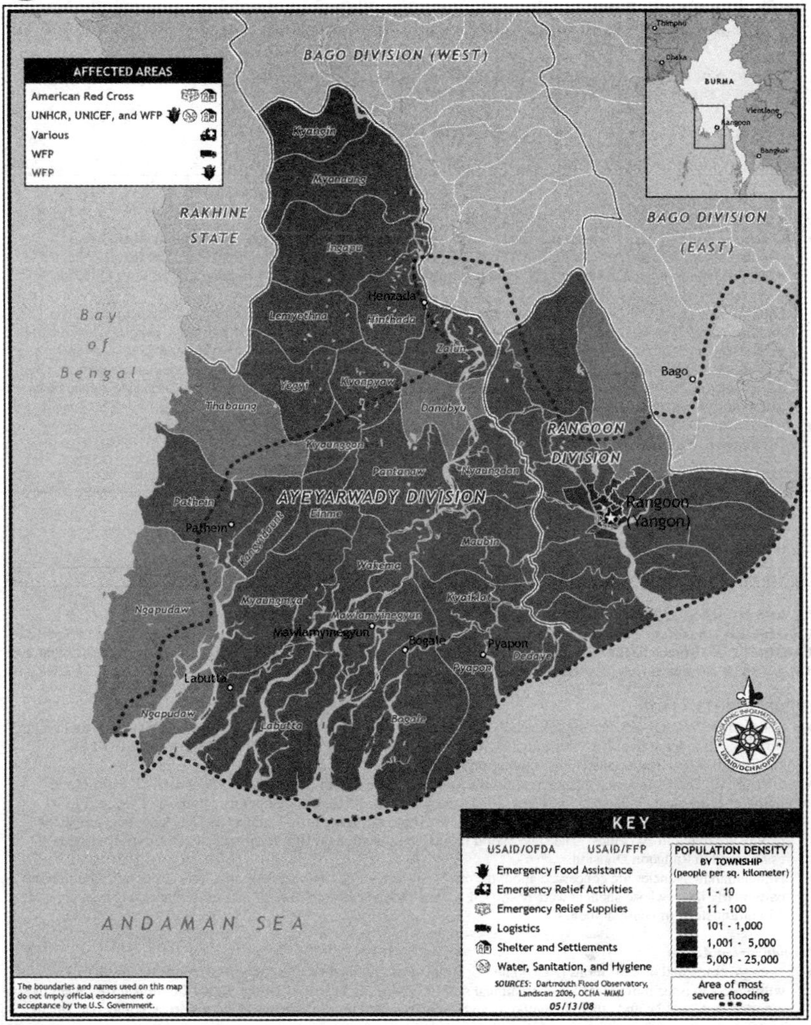

USAID
FROM THE AMERICAN PEOPLE

BUREAU FOR DEMOCRACY, CONFLICT, AND HUMANITARIAN ASSISTANCE (DCHA)
OFFICE OF U.S. FOREIGN DISASTER ASSISTANCE (OFDA)

Burma – Cyclone

Fact Sheet #6, Fiscal Year (FY) 2008 May 13, 2008

Note: The last fact sheet was dated May 12, 2008.

KEY DEVELOPMENTS

- On May 13, the USAID Disaster Assistance Response Team (USAID/DART) and U.S. Department of Defense (DOD) coordinated two airlifts of USAID commodities from Utapao, Thailand, to Rangoon. The flights delivered 4,200 insecticide-treated mosquito nets, 12,680 bottles of drinking water, 4,400 blankets, and 120 rolls of plastic sheeting, valued at $107,690.
- On May 12 and 13, two USAID/OFDA-contracted planes transported 129 metric tons (MT) of commodities from regional warehouses in Dubai, United Arab Emirates, to Utapao for onward delivery to Burma, pending Government of Burma (GOB) approval. The supplies include 1,100 rolls of plastic sheeting to serve an estimated 55,000 people; 19,600 hygiene kits, which can serve 98,000 people for two weeks; 25,200 ten-liter water containers to serve approximately 50,400 people; two 10,000-liter water bladders for water distribution; and two inflatable zodiac boats and motors to assist with relief operations.
- The USAID/DART remains in Bangkok, Thailand, coordinating U.S. Government (USG) assistance and awaiting GOB approval to enter Burma.

ESTIMATED NUMBERS AT A GLANCE		SOURCE
Total Dead	34,200 63,000-101,000	GOB – May 13, 2008 OCHA[1] – May 9, 2008
Total Missing	27,800 220,000	GOB – May 13, 2008 OCHA – May 9, 2008
Internally Displaced Persons (IDPs)	206,000 (in camps)	OCHA – May 10, 2008

FY 2008 HUMANITARIAN FUNDING PROVIDED TO DATE
USAID/OFDA Assistance to Burma $5,508,346
USAID/FFP Assistance to Burma $12,000,000
Total USAID Humanitarian Assistance to Burma $17,508,346

CURRENT SITUATION

- The cyclone severely affected approximately 1.2 million to 1.9 million people, and cyclone-related deaths could range from 63,000 to more than 101,000, with 220,000 people reported missing, according to OCHA. The GOB has confirmed 34,000 deaths and 27,800 missing persons, media sources reported.
- On May 10, OCHA reported that approximately 206,000 people in four townships of Ayeyarwady Division were living in 218 makeshift camps. In Bogale, 20,000 people have settled in 50 temporary camps. In Pyapon, 16,000 people are residing in 31 camps. In Labutta, 150,000 people are living in 117 camps. In Mawlamyinegyun, 20,000 people are living in 20 camps. The U.N. World Health Organization (WHO) reported that 640 camps have been established in Rangoon Division.
- Humanitarian agencies are concerned about the effects of heavy rains and harsh weather on cyclone-affected families, particularly those whose shelters were destroyed. Total rainfall between May 12 and 19 is expected to be more than 150 millimeters, or nearly 6 inches.

Humanitarian Access

- Non-governmental organizations (NGOs) and U.N. agencies continue to receive varying reports on the GOB's willingness to issue more visas for international staff. On May 12, U.N. Emergency Relief Coordinator John Holmes reported that 34 U.N. and other humanitarian staff were expected to receive visas to enter Burma, noting that "many more" staff were still awaiting visas.

[1] U.N. Office for the Coordination of Humanitarian Affairs (OCHA)

Logistics

- The U.N. World Food Program (WFP) and logistics cluster members are investigating possibilities of transporting commodities to affected areas by small trucks, boats, and helicopters. Logistics cluster meeting participants reported that the Rangoon port reopened on a limited basis on the morning of May 12.

Water, Sanitation, and Hygiene

- The U.N. Children's Fund (UNICEF) is re-commissioning an emergency water treatment plant used during the tsunami response to be deployed to Pathein township, Ayeyarwady Division, where many IDPs reportedly are gathering. Two NGOs transported eight small portable water treatment plants and 1.4 million water purification tablets to Rangoon on May 12. UNICEF has requested 23 large water bladders from the organization's stocks.

Food Security

- As of May 12, WFP had dispatched a total of 361 MT of food assistance to affected areas. Of this total, WFP has distributed 175 MT, assisting approximately 28,400 people.

Health and Nutrition

- Medical staff continue to treat cyclone-related injuries and focus on the prevention of communicable diseases such as diarrhea, acute respiratory infections, measles, and dengue fever, according to WHO. No outbreaks of malaria have been reported to date, according to aid agencies. WHO plans to share initial assessment results in the coming days.
- WHO confirmed the delivery of eight emergency health kits, each of which can serve 10,000 people for three months. As of May 12, WHO had delivered two of the health kits to Bogale and Labutta townships, and planned to dispatch four kits to Pyapon, Ngapudaw, Myaungmya, and Maubin townships in Ayeyarwady Division.
- As of May 13, WHO has dispatched 30,000 surgical masks, 30,000 gloves, and body bags for the collection of dead bodies in Bogale and Labutta townships. In addition, WHO has provided guidelines for the management of cadavers.

USG HUMANITARIAN ASSISTANCE

- On May 5, U.S. Chargé d'Affaires Shari Villarosa declared a disaster in Burma due to the effects of Cyclone Nargis. In response, USAID/OFDA deployed a USAID/DART and immediately provided $250,000 to UNICEF, WFP, and the Office of the U.N. High Commissioner for Refugees (UNHCR) for emergency food, water and sanitation, and shelter assistance.
- On May 6, an additional $3 million from USAID/OFDA was allocated for the provision of emergency relief assistance to cyclone-affected populations, including $1 million to the American Red Cross and $2 million for NGO partners.
- On May 12, USAID Administrator Fore announced $13 million in food aid and logistics assistance through WFP.
- On May 12 and 13, the USAID/DART and DOD coordinated the delivery of more than $175,000 of USAID/OFDA commodities to Rangoon on three DOD C-130 flights.
- U.S. Navy ships currently remain in the vicinity of the affected areas of Burma for training exercises and could be redirected, if necessary to support relief efforts. The U.S. Military continues to make plans to support potential relief operations to Burma.

USAID HUMANITARIAN ASSISTANCE TO BURMA

Implementing Partner	*Activity*	*Location*	*Amount*
	USAID/OFDA ASSISTANCE[1]		
American Red Cross	Emergency Relief Supplies; Shelter Kits	Affected Areas	$1,000,000
Various (TBD)	Emergency Relief Activities	Affected Areas	$2,000,000
UNICEF, WFP, and UNHCR	Emergency Food Assistance; Shelter and Settlements; Water, Sanitation, and Hygiene	Affected Areas	$250,000
WFP	Logistics	Affected Areas	$1,000,000
Various	Emergency Relief Supplies	Affected Areas	$1.258,346
TOTAL USAID/OFDA			**$5,508,346**
	USAID/FFP ASSISTANCE[2]		
WFP	P.L. 480 Title II Emergency Food Assistance	Affected Areas	$12,000,000

TOTAL USAID/FFP	**$12,000,000**
TOTAL USAID HUMANITARIAN ASSISTANCE TO BURMA IN FY 2008	**$17,508,346**

[1] USAID/OFDA funding represents anticipated or actual obligated amounts as of May 13, 2008.
[2] Estimated value of food assistance.

PUBLIC DONATION INFORMATION

- The most effective way people can assist relief efforts is by making cash contributions to humanitarian organizations that are conducting relief operations. Information on identifying such organizations is available in the "How Can I Help" section of www.usaid.gov – Keyword: Cyclone Nargis, or by calling The Center for International Disaster Information (CIDI) at 703-276-1914.
- USAID encourages cash donations because they allow aid professionals to procure the exact items needed (often in the affected region); reduce the burden on scarce resources (such as transportation routes, staff time, warehouse space, etc); can be transferred very quickly and without transportation costs; support the economy of the disaster-stricken region; and ensure culturally, dietary, and environmentally appropriate assistance.
- More information can be found at:
  - USAID: www.usaid.gov – Keyword: Donations
  - The Center for International Disaster Information: www.cidi.org or (703) 276-1914
  - Information on relief activities of the humanitarian community can be found at www.reliefweb.int

USAID/OFDA bulletins appear on the USAID web site at http://www.usaid.gov/our_work/humanitarian_assistance/disaster_assistance/

For the record: Sein Win, prime minister in exile of Myanmar

Premier in exile: the Burmese will rise up

Sein Win

Age: 63
Born: Taungdwingyi, Burma
Lives: Maryland, United States
Family: Wife, Si Si Thwe, two children
Background: Doctorate in mathematics from Hamburg University (Germany). Professor of mathematics at Colombo University (Sri Lanka) and Nairobi University (Kenya)
In the news: Leader of the Burmese government in exile

INTERVIEW. Myanmar has a ruling military junta – and then there's Sein Win, prime minister in exile. In 1990 Myanmar held democratic elections, but the junta ignored the opposition's overwhelming victory. Democracy champion Aung San Suu Kyi was put under house arrest. So exiled politicians formed a coalition government, led by Sein Win. This soft-spoken former math professor, who is also Aung San Suu Kyi's first cousin, has since been re-elected prime minister three times. He lives in exile in the United States.

Shortly after a cyclone struck Myanmar in the beginning of May, the secretive military regime held a referendum on a new constitution that excludes many opposition politicians from running for office. The junta claims that 92% of the population voted in favor of the constitution, but human rights organizations report that the voting was rigged.

Is outside humanitarian aid reaching the thousands of cyclone victims in Burma?
A lot of supplies are piling up, but the military picks the places where they want to send the aid. There is enough aid, but it's just waiting because the regime doesn't let the aid organizations enter.

Has the U.N. put enough pressure on the leadership?
The U.N. is as strong as its member states. Strong countries like the United States, Great Britain, Germany, France and Italy should form a coalition to put pressure on the regime.

What can private citizens do to help the victims?
Beg your governments to put pressure on the Burmese government.

Have you heard from Aung San Suu Kyi?
Because she's held incommunicado I don't have any direct contact with her. I could go to Burma myself, but the regime wouldn't let me leave again. They'd like to have Aung San

Suu Kyi out of the country and me inside. As long as she's there they are afraid.

What is your reaction to the constitutional referendum?
How can you have a real referendum when large parts of the country are under water? It's very undemocratic. It excludes people who are married to foreigners from holding public office. This is clearly aimed at Aung San Suu Kyi. (Her late husband was the son of a British father and a Canadian mother.)

Last fall Myanmar had its Saffron Revolution. How long do you think the regime will stay in power?
We can't say an exact time, of course, but the regime is near its end. Its behavior is a crime against humanity. But I know that the Burmese won't stay quiet. They will rise up.